Amazing Flights and Flyers

Other Books by Shirlee Smith Matheson

Nonfiction

Youngblood of the Peace
This Was Our Valley
Flying the Frontiers Vol. I "A Half-Million Hours of Aviation Adventure"
Flying the Frontiers Vol. II "More Hours of Aviation Adventure!"
Flying the Frontiers Vol. III "Aviation Adventures Around the World"
A Western Welcome to the World (history of Calgary International Airport)
Lost – "True Stories of Canadian Aviation Tragedies"
Maverick in the Sky: The Aerial Adventures of WWI Flying Ace Freddie McCall

Teen Fiction

Prairie Pictures
City Pictures
Flying Ghosts
The Gambler's Daughter
Keeper of the Mountains
Fastback Beach
Jailbird Kid

Amazing Flights and Flyers

Shirlee Smith Matheson

Shirlee Smith Matheson

Frontenac House
Calgary, Alberta

Book and cover design: Epix Design
Cover image: Dan Ryan
Author photo: Paula Sten

Library and Archives Canada Cataloguing in Publication

Matheson, Shirlee Smith
Amazing flights and flyers / Shirlee Smith Matheson.

Includes bibliographical references and index.
ISBN 978-1-897181-29-4

1. Aeronautics--Canada--History. I. Title.

TL523.M365 2010 629.130971 C2009-905299-7

The publisher acknowledges the support of the Canada Council for the Arts for our publishing program. We also acknowledge the support of The Alberta Foundation for the Arts.

Canada Council for the Arts
Conseil des Arts du Canada

Alberta Foundation for the Arts

Printed and bound in Canada

Published by Frontenac House Ltd.
1138 Frontenac Avenue S.W.
Calgary, Alberta, T2T 1B6, Canada
Tel: 403-245-2491 Fax: 403-245-2380
editor@frontenachouse.com www.frontenachouse.com

Acknowledgements

Most goals need an assist, and books such as this could not be completed without the help of others whose interest in the topics mirrors my own.

I offer sincere thanks to those whose stories and photographs appear between these covers – who willingly related their own life stories and shared knowledge of significant adventures of relatives, friends and colleagues.

Secondly, I wish to thank Clark Seaborn, P.Eng., who once again checked out myriad technical aspects, as he has done with all my previous aviation books.

Contents

Maps and Headlines

Note on Sources and Dialogue

All dialogue and other speech in *Amazing Flights and Flyers* has been taken verbatim from newspaper accounts published at the time of the respective events; from internet accounts published either at the time or at a later date; from books, journals and other sources published at a later date; and from interviews conducted by the author either in person or by telephone or email with individuals who participated in or witnessed the events. Every possible effort has been made to confirm the correctness of such dialogue; in no instance is dialogue inferred or imagined as having been spoken by persons involved in the events.

Note on Units of Measurement

At a conference of the International Civil Aviation Organization, held in Chicago in 1944, a resolution was passed to use the metric system (Système Internationale d' Unités, or SI) as the primary international standard of measurement. That sounds straightforward – but it wasn't. As the ICAO itself commented in a later publication, "It was evident from the beginning that the achievement of standardization of units of measurements would not be easy."

Amendment 13 to the resolution, adopted in 1979, "recognized ... there are some non-SI units which have a special place in aviation and which will have to be retained .... These are the nautical mile and the knot, as well as the foot when it is used in the measurement of altitude, elevation or height only."

This diversity of standards creates vexing dilemmas for anyone writing about aviation. In *Amazing Flights and Flyers,* units of measurement are those that would have been used by the participants in the events described in the book. Thus for all events prior to the mid-1970s, when metric units were formally introduced in Canada, measurements are expressed in miles and feet rather than kilometres and metres; Imperial gallons rather than litres; tons, pounds and ounces rather than tonnes, kilograms and grams; and degrees Fahrenheit rather than Celsius. Exceptions are distances and weights used by the German navy during the Second World War which, as would be expected, used metric units (see Chapter 5, "Nazi Toads and other Wartime Invasions"). Liquid volumes (e.g., fuel capacity) described in Chapters 14 and 15, "Kenn Borek Air – 'Anywhere! Anytime'" and "So Others May Live" respectively, are expressed in gallons (in specified cases, U.S. gallons rather than Imperial gallons) because these units are the ones that the persons involved were accustomed to using. And – following ICAO practice – nautical miles and feet are generally used wherever appropriate.

Preface

My previous collections include *Flying the Frontiers (Vols. I, II and III)*, as well as *Lost –True Stories of Canadian Aviation Tragedies,* which chronicle the lives of people whose aviation adventures have amazed and inspired many readers. A number of those whose biographies appeared in these collections are now deceased, and their stories might simply have been relegated to old logbooks and archival documents. Through personal interviews and secondary research I recorded not only their stories but their personal insights and patterns of speech as they related the unbounded challenges of flying. I am grateful that I was able to capture these personal stories in time, as all are significant to our knowledge of Canadian aviation history.

There is always joy as well as sadness in lives that are lived to the fullest degree. The stories related in the aforementioned collections, as well as in *Amazing Flights and Flyers,* are no exception.

Some accomplishments seem to be beyond human endurance, such as the two mid-winter medical evacuation flights pioneered by the intrepid crew of Kenn Borek Air; the continuing efforts by volunteers from CASARA to search for lost people and planes; the determination of aviation pioneers who fight to fly the volatile conditions experienced in our Maritime provinces; the amazing lifestyles of those who choose to live in the Far North and never want to leave.

On the other side of the flying field are those who used flight as an opportunity for personal escapes or hijacking capers, or whose fates were suddenly decided by bad luck – engine failure, sudden weather changes, or chances taken with unfamiliar machines and terrain. A high-risk wartime story chronicles the attempt of enemy forces to dock their U-boat on Canadian soil to install a weather reporting station. All are remarkable stories, and most are little-known.

Flight can be a combination of thrills beyond compare – and sudden full stops. The stories in *Amazing Flights and Flyers* encapsulate nearly every human emotion and scenario, and range from the early days of the 20th century to the present.

A quote from David Henry Thoreau captures my feelings about chronicling all sides of the flying field:

> *Even the death of friends will inspire us as much as their lives …*
> *Their memories will be encrusted over with sublime and pleasing thoughts,*
> *As monuments of other men are overgrown with moss;*
> *for our friends have no place in the graveyard.*
>
> *("Though All the Fates Should Prove Unkind",*
> *A Week on the Concord and Merrimack Rivers - 1849)*

I hope readers will enjoy the trip – and cheer at the courage while marvelling at the audacity of these flights and fights experienced by our fellow adventurers.

Shirlee Smith Matheson

Deadmen's Valley

Deadmen's Valley, also known by the gruesome epithet "Headless Valley", is a beautiful area in the Northwest Territories near the junction of the South Nahanni River with the Liard River. As the crow – or an aircraft – flies, it's approximately 350 miles southwest of Yellowknife, Northwest Territories, NT, or 625 miles northwest of Edmonton.

The Nahanni has always been considered a supernatural place by the aboriginal people living there. But it gained the attention of non-aboriginals in the early 1900s when rumours of a Northern tropical vale warmed by Chinook winds, natural hot-springs and gold-filled streams brought explorers from "the outside". Then came discoveries of at least 25 mysterious disappearances or deaths, including several headless corpses. Whether the work of unscrupulous travellers, wilderness shamen, Windigo spirits, or natural phenomena, the legends have only enhanced the Nahanni's macabre reputation.

The McLeods' Misfortune

In the late 1800s, brothers Willie, Frank, and Charlie McLeod learned of the Nahanni's gold from natives' stories told, and nuggets shown, to their father, Murdoch, the Hudson's Bay factor at Fort Liard. The McLeod boys never forgot these tales, and in January 1904 they tried their luck, but their panning and sluicing brought few rewards. Their inaugural trip through the Flat River canyon took them over The Cascades of the Thirteen Drops to the South Nahanni, where their boat was swamped – a warning of future disasters that the McLeod brothers failed to heed.

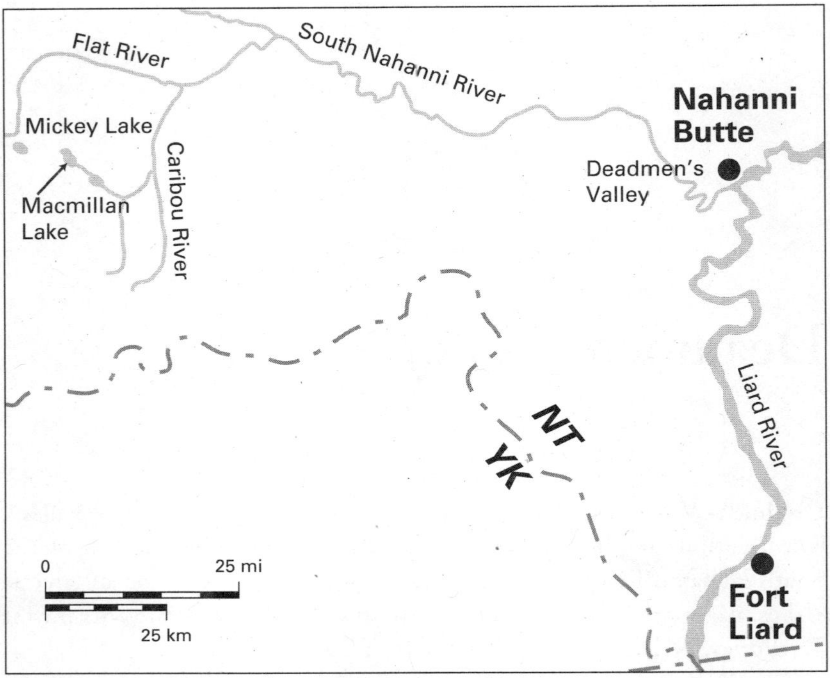

Deadmen's Valley and the Nahanni

The following year Willie and Frank, with Scottish engineer Robert Weir, returned to seek the elusive mother lode. When they didn't come back, their brother Charlie sent out a search party. In a steep valley on the left bank of the South Nahanni River they found what remained of Willie and Frank: two sets of arm bones, two pairs of feet, and one lower jaw. They also found a note scratched on wood: "We have found a fine prospect."

Then police were told of a partially complete skeleton found that past fall on the headwaters of a tributary to the South Nahanni. What or who could have killed the two brothers and the third person, thought to be their partner Robert Weir, remains unsolved.

On September 28, 1915, prospectors found a skeleton 70 miles upriver from the McLeods' resting place – again, headless – but identified as prospector Martin Jorgensen. Then, in 1922, John O'Brian, a former soldier turned trapper, was found frozen to death in the Nahanni.

In 1927, the body of Charles Taylor, also known as "Yukon" Fisher, was reported by a "Captain" Knox near the headwaters of the Flat River, a tributary to the South Nahanni between Virginia Falls and what later became known as Headless Valley. His cabin was burned to the ground. While no one had previously seen Taylor in this area, he had sold furs at the trading post at Ross River, and paid for supplies with coarse gold nuggets he said he'd acquired from prospecting.

In 1929, Angus Hall left his two gold prospecting partners to start work on a cabin at Bennett Creek. He disappeared from their sight over a hill and was never again seen.

The next unexplained death was trapper/prospector Phil Powers, whose remains were found in the fall of 1932, in a heap of ashes on his bunk bed inside his main cabin. Nearby was a rifle with a wire attached to the trigger, known to trappers as a "set-gun". In an odd juxtaposition, his crushed skull lay near his foot bones. A pencilled note nailed to a post on his food cache said, "*Finis - 1932 - Phil Powers.*"

The RCMP now had eight unsolved deaths on their records. In addition were five reports of disappearances near the headwaters of the Macmillan River and on the Yukon River, where no remains were ever found. Again, a number of these accidental deaths had mysterious aspects: one, in 1928, involved a canoe carrying three passengers, a physician from Dawson City, two companions and a dog. Only the dog was found, alive, on the river-bank. When the canoe was located an inspection showed a hole had been punctured in it, *from the inside out.* While few of these deaths (or presumed deaths) were recorded in government records in the Yukon or Northwest Territories, they have been substantiated with clarity by local residents.

The Mad Trapper

Aviation brought the entire North Country into closer scrutiny in the early 1930s, most famously through the story of The Mad Trapper of Rat River.

On Christmas Day, December 25, 1931, a complaint was made to RCMP Constable Millen at the Arctic Red River post that a man going by the name of Albert Johnson was springing neighbouring traplines and hanging the traps up in trees. The next day, Millen ordered Constable Alfred King and Special Constable Joe Bernard to check it out. This meant that the officers would have to set out in -40° F weather with two dog teams to resolve the situation with Johnson at his remote cabin on the Rat River.

They stayed overnight at Fort McPherson, 30 miles in to their journey, and then carried on, again in bitter cold weather, to travel the remaining 25 miles. They made an overnight bush camp at the mouth of the Rat River, and in the morning continued up the Rat to meet with Johnson, arriving at his small spruce-log cabin around noon on December 28.

Shouting a greeting from outside in the Northern custom, and receiving no answer, although smoke trailed from the chimney, Constable King stepped up to the small, four-foot-high door and knocked. Johnson was seen silently staring out through a burlap sack that covered a tiny window, but he made no reply.

Requiring a search warrant to proceed further, King and Bernard retreated, mushing down to Aklavik to tender their report to Inspector Eames, commander of the Arctic sub-region.

With reinforcements of two more men, Constable R.G. McDowell and an aboriginal Special Constable, Lazarus Sittichinli, and carrying the necessary search warrant, they left on December 30 to make a return visit to Johnson's cabin. The scenario was repeated: although Johnson was inside his home he did not respond to their calls or knocks. As Constable King approached the door from the side farthest from the window, a shot shattered the silence of the frigid winter day. King crawled away through the snow toward protection on the riverbank. The officers then fired shots into the cabin, which Johnson returned fire for fire.

Realizing that King was severely injured, the main object for Sittichinli was to transport him back the 80 miles to Aklavik to receive medical aid. It took 20 hours, and King's life was saved – the bullet had penetrated and passed through his chest but had not struck vital organs.

A Second Visit to Rat River

On January 4, 1932, a party comprised of seven men (Inspector Eames, Constable McDowell, Special Constables Joe Bernard and Sittichinli, and civilians Ernest Sutherland, Karl Gardlund and Knut Lang), along with 42 dogs, left from Aklavik. The Royal Canadian Corps of Signals station in Aklavik sent word to Constable Edgar Millen (who claimed he'd once previously met Johnson) and an aboriginal guide, Charley Rat, to join the party. They rendezvoused at Blake's Post, where 20 pounds of dynamite was added to their arsenal.

Eight miles from Johnson's cabin they made camp, and there planned how best to approach this obviously dangerous man. Dealing with extreme cold weather and short hours of Arctic winter sunlight, they strategized manoeuvres and travelled around the territory to familiarize themselves with back trails and areas where they could possibly be ambushed by Johnson.

Calls to Johnson to come out of his cabin continued to be ignored, and meanwhile the food supply for the men and the dogs was quickly being diminished. The only choice to get Johnson out quickly was to blast him from his hole. The explosion blew off part of the cabin roof and knocked off the stovepipe, but still Johnson refused to give himself up.

Shots continued to be fired uselessly from both sides. Finally, in the middle of the night of January 10, Inspector Eames ordered that the remaining four pounds of dynamite sticks be bound together and hurled at the cabin. The blast completely ripped off the roof and knocked in one wall. Silence. Expecting the man to be dead, Karl Gardlund was recruited to creep toward the cabin with a flashlight to see if he could spot the man's body. Suddenly a very alive Johnson fired, knocking the flashlight clean out of Gardlund's hand.

Now determined to take this mystery man alive rather than set the cabin on fire, Inspector Eames ordered the posse to return to Aklavik.

Third and Fourth Visits to Rat River

On January 14, 1932, Constable Millen and Karl Gardlund returned to Rat River, to find that Johnson had fled. This was not unexpected, but he was a dangerous man and now his whereabouts were completely unknown.

The news of his standoff had now reached the media, with the name "The Mad Trapper of Rat River" being applied to the story, which was far from over.

A second posse was formed on January 16, comprised of ex-RCMP officer John Parsons, trappers Frank Carmichael, Noel Verville and Ernest Sutherland and Special Constable Lazarus Sittichinli, along with two men from the Signals Corps, Staff Sergeant Earl F. Hersey and Quartermaster Sergeant R.F. Riddell. A radio was added to their equipment to relay messages.

They set out to track Johnson up the Rat River in the -47° F weather and long hours of Arctic darkness. Again the food supply was quickly diminished to feed the large group of men and dogs, and a request was made for Staff Sergeant Hersey and an aboriginal man to bring more supplies from Aklavik.

The posse's desperate search for signs on the trail was rewarded on January 30 when Johnson's camp was spotted in a canyon far up the Rat River. Constable Millen and Noel Verville advanced down toward the frozen river, as Riddell and Gardlund descended into the canyon from another position.

Suddenly a shot blasted from Johnson's 30-30 Savage. Millen and Verville dropped to the ground to return fire. Silence. Was he dead? If not, how long could everyone lie in wait in the snow? Two hours passed. Against Verville's advice, Millen made a plan to go in after him, with Riddell joining him, and covered by Gardlund and Verville. Spotting movement, Millen fired, and Johnson quickly returned fire. Both shots missed their targets. Millen fired once more, and Johnson responded with two bullets in rapid succession. Millen fell, shot through the heart.

As the posse retreated to Aklavik with their dead companion, the news media fed details to the public. People expressed horror on hearing of this drama, but were also amazed by the incredible survival skills of this enigmatic fugitive, and excitedly followed newspaper and radio reports that told of something else besides The Great Depression.

On February 2, 1932 a posse was formed once again to track the now famous Mad Trapper of Rat River. On arrival at the scene of the previous shoot-out, it was discovered that Johnson had escaped the canyon by chopping hand-holds in the snow and ice with his axe and climbing up the vertical cliff behind him. He had now fled, to hike over the 6,000-foot range of the Richardson Mountains, in temperatures as low as -70° F, and disappear into the vastness of the Yukon.

Surveillance by Air

In desperation, Inspector Eames issued an unprecedented request: could an aircraft come to their assistance? Edmonton's famous First World War flyer Wilfred "Wop" May and mechanic Jack Bowen responded. On February 5 they flew in to Aklavik via the Mackenzie River in a Bellanca CH-300 Pacemaker (CF-AKI) monoplane to transport RCMP members and essential supplies, and provide aerial surveillance, across the Northwest Territories and into the Yukon.

Johnson had now been living off the land for 30 days, snowshoeing through an extremely cold, snow-covered, and seemingly barren land. What could be sustaining him? He couldn't shoot large game – much too unwieldy to butcher and pack – and the shot would be heard by his pursuers. There was very little, if any, edible vegetation available. Could he live by snaring rabbits or squirrels, catching mice? He'd have to eat them cold and uncooked, for he could not build a fire for fear of it being spotted. It seemed impossible. But, he was still alive, still running, and baffling the intelligence, stamina, and equipment of modern civilization.

Wop May and Jack Bowen were battling the elements as well, trying desperately to keep the Bellanca in the air in the extreme cold, snow and fog that beleaguered flight and at times kept the aircraft grounded.

Constable Sidney May (no relation to Wop May), Special Constable John Moses, and several trappers were added to the posse on February 8, after travelling by dog team over the Richardson Mountains to the headwaters of the Barrier River, to monitor known trails through the passes. When a message came on February 12 that strange snowshoe tracks had been seen east of Lapierre House (a Hudson's Bay post at the junction of the Bell and Water rivers, an important stop between Fort McPherson and Fort Yukon), Constable May travelled back to the camp at the mouth of the Rat River to inform the rest of the party. Incredibly, Johnson's tracks were picked up near the headwaters of the Barrier River on the far side of the Richardson Mountains, moving along a creek to the Bell River southwest of Lapierre House, and then south on the adjoining Eagle River.

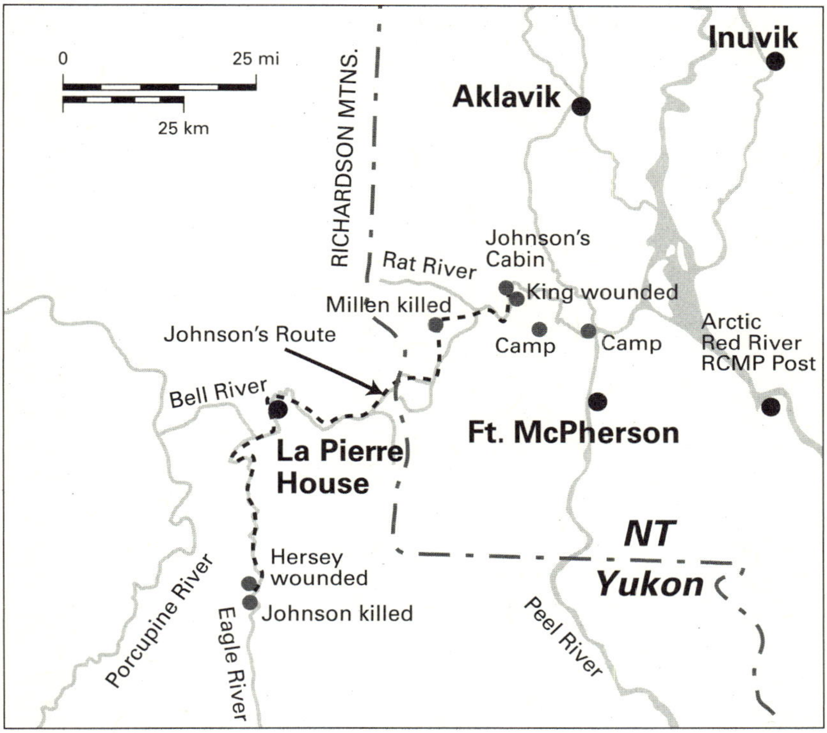

The Trail of Albert Johnson

On February 17, 1932 he was spotted cleverly backtracking on his snowshoes (stepping into his old tracks to fool his pursuers as to his direction), believing the posse to be behind him. Instead, they were in front. Sergeant Hersey was the first to encounter him. As the two men momentarily stared at one another, Johnson quickly donned his weighty 10-pound snowshoes and started toward a hairpin turn on the east bank of the river. Hersey fired, joined by Verville and Constable Sid May. Johnson ran, then whirled and fired, hitting the kneeling Hersey with a shot that ripped through his elbow, knee and chest.

Johnson then changed directions, throwing himself down into the snow in the middle of the frozen river, and hunkering down behind his backpack in preparation for the shootout, his rifle pointing downstream. Three men from the posse fired at him from the west bank, while two others lay firing at him from the snow-covered surface of the middle of the river. Seven bullets hit the fugitive, striking him in his shoulder, in his side, and one striking the ammunition he had in his pocket with such force that the explosion ripped a piece of flesh from his thigh. Still Johnson kept firing, until a final and fatal bullet through his back finished his life.

He had led Mounties, and local people, both white and aboriginal, who were completely familiar with the bush, and aided by aircraft, on a nearly 400-mile chase lasting over 54 days in unbelievably rugged conditions, one man against many, and never surrendering until a hail of bullets ended his flight.

Investigations

What is The Mad Trapper's connection to the mysteries of the Nahanni? When the body of Albert Johnson was examined, his pockets and pack contained basic survival supplies, armaments (a .30-30 Savage rifle, .22 calibre rifle and a 16-gauge sawed-off shotgun) and a huge stock of ammunition, fishhooks, cash (including two American $5 bills), five pieces of gold tooth fillings (not part of his own dental work), some alluvial gold, and five pearls.

Rumours persist to this day that some of these items could connect him to unsolved deaths in the Nahanni that had occurred subsequent to his arrival in the area in 1927. In fact, he might actually have been a man called Arthur Nelson who came from the United States, and lived 40 miles from the headwaters of the South Nahanni and often visited the regions of the upper Macmillan River.

Dick North, author of *The Mad Trapper of Rat River*, provides this plausible connection:

> *If Johnson and Nelson were the same person, and if Nelson was a killer and was stealing his victims' teeth, he would naturally remove the head [to lessen chances for suspicion if the skulls were found lacking gold teeth]. He would also know that any man murdered in such a wilderness would have his body picked clean by ravens, camp robbers and assorted flesh-eating mammals within a matter of weeks. This would adequately disguise any evidence of wrongdoing.*

Therefore, says North, the mysterious deaths of Charles Taylor, Angus Hall, and Phil Powers in this region might possibly be traced to Johnson/Nelson. Further investigative work by Dick North has led him also to suspect that Albert Johnson might have been a man from North Dakota named Johnny Johnson, born in Norway in 1898, and said to be a bank robber and a horse thief.

Technology had made some headway, but whether it could resolve all these mysteries still remained just that, a mystery.

Digging up Bones

Dick North tried a number of times to gain permission to exhume the grave in Aklavik of Albert Johnson, but the requests were always denied. However, 15 years later, in 2006, a team from the University of Alberta (comprised of anthropologist Owen Beattie; forensic investigator Lynne Bell; Carrie Gour, a past resident of Inuvik, a village in the Yukon near Aklavik, and now a producer for Alberta's "Myth Merchant" films; and Emmy award-winning filmmaker Michael Jorgensen), came armed with a contract from Discovery Channel for a documentary film (which, however, was contingent upon winning elusive permissions from elders in the Gwich'in community of Aklavik). Included later in the committee were John Evans of the Canadian Police Research Centre, who would take three-dimensional readings of the skull; and experts in the fields of DNA, pathology, and forensic dentistry.

Years had passed since the drama and death of The Mad Trapper, and the identity of "Albert Johnson" would surely mean something to his family, reasoned Knute Hansen, Mayor of Aklavik. Through an introduction made by Mel Benson, business and cultural guide for the project, Carrie Gour tentatively approached Charlie Furlong, Chief of the Aklavik Band, for permission to make the documentary in Aklavik – and exhume the body so it could be subjected to the most modern testing.

All involved knew it was a highly sensitive issue, as many prior attempts had been turned down flat, but Gour's approach to work with and for the community, and champion the contribution made by aboriginal guides during the 1932 manhunt, ultimately met with approval. Dennis Allen, an Inuvik resident and filmmaker, was assigned as the community liaison.

Further permissions for the exhumation were sought from, and granted by, Reverend Andrew Atagotaaluk, Anglican Bishop of the Arctic, as well as the Registrar General of the Northwest Territories, who is in charge of the cemetery grounds.

An ironic note is that because Albert Johnson was a murderer he could not be buried on sanctified ground, so his gravesite had been placed outside the fence. However, over time the graveyard expanded, and so the outlying area where Johnson's bones reposed became the property of the church. Now, Johnson's grave lies within the hallowed grounds of the little cemetery in Aklavik, and its upkeep comes under the auspices of the church.

And so permission was finally granted from all the protectors of this sanctified site to exhume, examine and film the contents of the grave of the mystery man known to the world as The Mad Trapper, with the hope of providing absolute identification.

The casket was opened and there rested Albert Johnson.

"I was blown away by my first glimpse of his face, and the recognition of his still-preserved turned-up cowlick of hair on his forehead," said Dr.

Beattie. "It appeared exactly as it did in the period photographs taken in the hours immediately before his burial."

The remains underwent DNA (mitochondrial) testing in a thoroughly scientific attempt to solve the mystery of the true identity of Albert Johnson, and comparisons were made with DNA samples provided by various persons who thought they might be related to the man known as The Mad Trapper. At the time of this publication (2009), however, no conclusion has been reached regarding Albert Johnson's genealogical lineage, and so the mystery endures.

Many years have passed since the mysterious disappearances or deaths of isolated trappers, or the Albert Johnson case, but the Nahanni's remoteness still provides a sanctuary for those who seek peace, and also protection for those who seek privacy.

In 1936, trappers Joseph Mulholland and Bill Epler disappeared without a trace in the Headless Valley. In February of that year, pilot George Dalziel had flown the two men in to Glacier Lake up the Nahanni. When they were 23 days overdue for the pre-arranged rendezvous, Dalziel flew in but could find no trace of the men, and their cabin was burned to the ground. The bodies were never found, but it is known that they did not perish in the fire. Could they have fallen through ice? Been trapped in a snow slide while attempting to cross the mountains near the junction of the Flat and South Nahanni rivers? Unknown.

These stories, and possibly others, continued to strengthen the unshakable legends. Quite understandably, the media featured a number of these incidences, and the Calgary *Albertan,* in an issue dated February 19, 1947, coined the name *Headless Valley,* and also referred to the South Nahanni as the *River of Mystery.* Similarly, an Alaska Highway Handbook deemed Fort Nelson to be "the jumping off point for *The Valley of Vanishing Men.*"

Whether the name draws or repels visitors is a moot point. Adventurers may love the beauty of the Nahanni and its storied history, but it's still a wild country, with many secrets known only to a few privileged keepers.

Pictures of the Nahanni

The stories of a Nahanni gold hunt gone bizarrely wrong were all over the newspapers and courts from May 1960 to May 1961. Five greenhorns who had been prospecting and sharing cabins on Macmillan Lake in Deadmen's Valley had apparently run out of food when their scheduled aircraft failed to pick them up. Two of the desperate men had gone in search of help and were never found. One took his own life by dynamite. Fortunately for the two remaining survivors, northern bush pilots Jim and Chuck McAvoy of Yellowknife happened to notice an SOS message tramped out on the soft ice of the lake and had rescued them in their Cessna 180. The "Starvation Cabin" episode was added to the legend of the Nahanni.

One person who heard the story firsthand from Jim McAvoy was Angus Blake MacKenzie, an Edmonton filmmaker and himself a pilot, with navigation training from the Second World War and 415 hours on his private pilot's licence. MacKenzie knew he had the basis of his next documentary film.

The film, *Nahanni – the Lure of Gold,* would focus on this story of the five unlucky prospectors at Starvation Cabin and would also delve into the macabre history of Deadmen's or Headless Valley, which had already been the scene of so many mysterious disappearances and deaths.

In late October of 1961, MacKenzie and John Langdon, a bush pilot and photographer for the film project, leased a Cessna 180 (CF-FYN) to check out the territory and also film some placer mining operations and general scenes in the Nahanni. When the snow fell, Langdon took the aircraft in to replace the floats with skis in preparation for the return to the Nahanni.

At this time the filmmakers met with Nahanni Butte residents Vera and Dick Turner, with whom they discussed their project and hinted at the possibility of finding rumoured lost gold. Dick Turner, himself a pilot, warned the filmmakers that, with the advent of aviation, the curse of the Nahanni had continued to add to its human toll. All northern pilots as well as engineers, trappers and miners could reel off numerous near-misses or crashes. But people still came, the Turners said with a sigh – by boat or aircraft – for adventure. Few left without a story, and some didn't survive to tell their tale.

MacKenzie's own story started immediately, when he and his partner were snowed in for two weeks. They were lucky to find shelter with the Turners. Then, in late December, MacKenzie and Langdon were stranded again, this time for six days at Macmillan Lake. They were eventually found by search aircraft after the storm abated, and were able to fly their aircraft back to Fort St. John. Unfortunately, the filmmakers heeded neither of those first warnings and continued with their project.

The next step was to hire two local prospectors, whose familiarity with the territory and the gold business would add authenticity to the film. Tom Haggerty was from the Yukon; Jack Mulholland had trapped and prospected in the Liard region from the 1930s. His brother Joe, as well as Jack's former trapping partner, Bill Epler, had both disappeared without a trace in 1936 in Headless Valley, which would lend further credibility to Jack's role in the film. In December, Langdon flew Haggerty and Mulholland north the 92 miles from the Smith River airport to the film crew's base camp at Mickey Lake, NT, one and one-half miles west of Bennett Creek.

On January 5, 1962, needing supplies, MacKenzie took off for the 50-minute flight to Smith River, near the ominously named Valley of Lost Planes. At the time of his return flight the ceiling between Smith River and Mickey Lake had dropped below 500 feet, and visibility was only one-half mile, with snow. But the pilot had continued on. He was never seen again.

When he did not complete his flight plan a massive search was organized, although it was unable to start until January 17 due to weather. Nothing was found after 40 days, although local searches continued when and where possible. Langdon carried on with photography for the film, with the help of the two other men.

On August 14, 1962, the wreckage of CF-FYN was spotted while shooting footage for the closing minutes of the film. The Cessna 180 lay nestled right-side-up amid a grove of spruce trees in a valley at the 4,500-foot altitude, just eight miles southeast of Mickey Lake. Signs of MacKenzie's camp were found a short distance away. Search and Rescue personnel of the Royal Canadian Air Force (RCAF) returned to make a ground search for MacKenzie, or his body, and take an inventory of his camp. They found 46 notches chopped into the bark of a nearby tree. Inside the aircraft were a chain saw, gas and oil, and enough food to last for 100 days.

As well, police found two diaries, the first written with a pen and the second with the sharpened head of a bullet. Both were weathered and the writing nearly obscured, but told a story that only added to the Nahanni's mysteries.

One of the first entries in MacKenzie's diaries explained the circumstances of the crash:

> *Both Clark and Macmillan lakes were visible as I headed north, although there was a pile of stuff north of the Cariboo which I decided to fly under. There were patches of fog but I could still see the ground. I made a climbing turn to clear rising ground but the aircraft struck a tall tree because of the down draft.*

The tree sheared off the left wing and sent the aircraft spinning to the ground.

MacKenzie clambered out and examined his injuries: two sprained thumbs and a wrenched back. Not bad. He walked 100 yards to a rise to check his location. His white aircraft, buried in the snow-and-tree-covered slope, would be almost invisible from the air. In order to see it, searchers needed to make a dangerous downward sweep facing into the sun, and they wouldn't even begin to search until after his due-date, one week away. He knew where he was, however. Just eight miles from camp. Not far, as the crow flies.

At first he probably remained confident that he'd soon be found, and set about to make the best of the situation. As he recorded in his journal, he fashioned a canvas shelter from the engine cover and brought out his bedroll. He had a two-burner wood stove, and emergency rations for five days plus the large grocery order intended for the camp. He had an axe, snowshoes, warm clothing, summer-weight sleeping bag, and two rifles with plenty of ammunition. While he waited he chopped wood, sent out periodical radio signals and ploughed trails through the waist-high snow.

In his journal, he expressed concern about his partners who were waiting at camp:

> *They'll need the supplies, so I have decided to make tracks as quickly as possible. I have sent off a series of Maydays on my aircraft radio and will do so again. I have made some trails in the open, headed on toward the mountain. But unfortunately cloud after cloud has obscured the valley since I got out of the aircraft so there is little chance that these trails will be seen.*

Snow continued to fall and the wind whipped it into high drifts. The temperature dropped to -50° F. It took MacKenzie 10 hours of trudging over a three-day period to make a trail up to the mountaintop. On January 8 he wrote:

> *It was brutal work, and it left me terribly weak. I couldn't complete the trip to Mickey Lake, so I'm beginning to have doubts about my ability to ever do so. But at least now I'm sure of my location. I could see Mickey Lake clearly from the mountaintop. I keep listening for aircraft engines, even though I know it's too soon for them to come. I heard voices, and then I discovered I was talking to myself.*

Each day, if the wind lessened, he would try again to hike to Mickey Lake. Not possible.

> *I made three-quarters of a mile in three hours, and I know now that I won't be able to make it. So I returned and fixed up camp for a long stay.*

By January 10 his spirits began to falter:

> *I'm too tired to leave camp. I haven't been sleeping well and it's beginning to tell on me. It tires me to chop down a couple of trees and up here you need 20 of the damned things a day.*

The journal entries in the first diary end here. Following a 10-day gap in communication, he continued to write in the back pages of his engine log book, using the .303 bullet.

On January 20, 1962, the temperature dropped to -65° F. This day he wrote a note to his daughter, Donna, who was attending the University of Alberta:

> *I have decided to enlarge them [the notes] a little so that if I don't get home you'll know how it was with me at the end. I'm afraid I leave you little else save grief.*

Several more attempts were made to hike out, once down the mountain toward Macmillan Lake 10 miles to the south over relatively flat ground to the site of "Starvation Cabin" where the five prospectors had undergone their ordeal two years earlier.

On January 21 he wrote:

> *My chances of being found are better if I get to the cabin, so I'm spending as much time trail-making as I can. It takes me about two hours to walk to the end of my trail and I spend three more pushing it ahead through heavy brush and five feet of snow. It is discouraging. I can walk back in 20 minutes over the last three hours' trail.*

At this point, the first search plane was sighted. MacKenzie quickly lit a signal flare. The plane flew on. Then flew over. *The noise reminds me of the Edmonton airport,* he wrote.

On January 27, he further noted:

> *The RCAF were west of here all day, but it was cloudy and I saw not a sign of an aircraft. I stuck close to camp and instead of putting my red sleeping bag out in the snow as I had been doing, I mounted it on a pole and tramped out a SOS in the snow in the open area near my camp. Now a sleeping bag, even a summer one like mine, is not light, but that wind stretched it out straight like a flag!*

January 28 brought more heartbreaking near-sightings:

> *I turned the radio on, sent a May Day with my position, and was heading down the valley when a big twin-engined bomber came roaring up the valley at about 100 feet. They'd heard me! The pilot was following my three-mile long, two-foot deep trail. My fire was belching white smoke. My sleeping bag was flapping in the breeze. I was saved! I stood there, waving my toque, an idiotic big grin on my face, and the bomber flew right over, lifted to clear the mountain ridge, and was gone. I went back and sobbed the story into the radio, but no one heard.*

He heard no more airplanes after the first week of February, and knew the search had been called off. In fact, it was terminated on February 10, after more than 300 hours of flying by both military and civilian aircraft.

Temperatures again dropped to the -60° F range. MacKenzie stayed in camp, keeping his lifesaving fire going and looking after his frostbitten feet, as his food supply dwindled. *I know I ate too much in the first two weeks, and have tried to make up for it since,* he wrote. He itemized what was left, then added, *so today I'll draw up a new diet sheet which I'll keep rigidly until I'm able to hunt and fish.*

Another airplane!

> *I heard an aircraft while I was taking inventory. It sounds like the plane I've seen earlier, going up Flat River way, probably to the tungsten mine [50 miles to the north]. He sometimes flies over Macmillan Lake. The sooner I get there and tramp out a big SOS in the snow, the better. Maybe I'll be seeing you before June. It's not snowing now and I need wood. So long for a while.*

The diary entries continue sporadically until the 20th of February.

What happened to this competent and supposedly bush-smart man remains a mystery. There was still a supply of groceries in the aircraft. The terrain was rough but not forbidding. To make his way on foot back to camp he could have hiked 2,000 feet up a gentle incline, walked four miles along the top ridge, then down through a forested area to Bennett Creek.

Steve Villers, a Fort Nelson pilot and owner of Villers Air Services Ltd., participated in the search and still finds some aspects puzzling. "At the time he took off from Fort Nelson we had a very cold spell, minus 40° Fahrenheit, with clear skies and not a cloud anywhere in the area. The snow was deep and it kept falling after MacKenzie crashed, making the aircraft's white paint difficult to spot.

"There was quite a big search with the RCAF using a Lancaster, and numerous private aircraft. The area was thoroughly covered where the aircraft was finally found."

However, he recalls that the sky stayed clear for more than a week, during which time steam and smoke from camps and oil rigs could be easily seen from over 50 miles away. While MacKenzie's diary entries noted that he had kept a fire going for many days, neither flames nor smoke from his fires were spotted by anyone in search aircraft.

"Strictly my own opinion," Steve Villers adds, "but I think, having crashed and been unhurt, he intentionally disappeared. Certainly no body or skeleton has ever been found, so your guess is as good as mine as to what happened in the end.

"I am sure that no one would intentionally crash an aircraft, but having crashed and not been hurt it would have been an ideal opportunity to disappear. It would have been difficult to walk out, but not impossible." Villers recalls rumours of some personal problems. And also rumours of someone spotting MacKenzie in Mexico a year later. "But these were stories that were going around, and were probably just that," Villers concludes, in a typically Northern "let it be" attitude.

But theories persist regarding the fate of Blake MacKenzie: he and Langdon had found gold, and MacKenzie escaped with the booty; MacKenzie hiked out to start a new life in Australia, or South America; he'd found a new partner, a woman. "Find the woman and you will find Blake," says Dick Turner in his book *Wings of the North,* while acknowledging that weather was a major factor in MacKenzie's fateful flight.

The RCMP declined to speculate, stating simply that the man was missing and presumed dead. His wife reported that he was a heavy sleeper and was subject to sleepwalking. Perhaps he'd wandered off. With a rifle and axe? Not likely. Friends believe he might have been the victim of an animal attack (although no signs indicated this, such as bits of bone or clothing being found). He could have fallen through the ice, and his body washed downstream in the spring-run off. But this would not take into account the disappearance of his snowshoes, axe and rifle, which were never found during searches of the ground or streams.

Northern bush pilot Jim McAvoy adds another twist to this tale. Blake MacKenzie had personally told Jim that he'd become intrigued with the story of the five prospectors and their plight at Starvation Cabin, which had happened just the year previous to MacKenzie's idea to go to the Nahanni and film his documentary. Could he also have known about the supposed nefarious plans of these prospectors to "salt" their mine with high-graded gold concentrate that Jim and Chuck McAvoy had spied in a box behind their cabin when they'd flown in and rescued the two survivors? Perhaps MacKenzie had heard of the box and its contents when he'd become "lost" in the area shortly after that story broke.

It is known that MacKenzie had withdrawn a sum of money from his bank account in Edmonton before leaving on this trip. "His disappearance had to be planned," states McAvoy, who knows the North thoroughly from a lifetime of flying throughout its most remote areas, and also from his family-owned northern mining ventures.

"He made no attempt to walk back to his camp. I think the diary was a fake, written and dated beforehand; no way had he sat there for 50 days writing in it, waiting to be found."

Some northern residents consider such explanation to be "common knowledge", but remain willing to let any judgements rest in peace.

Alfred North Whitehead, in his book *Adventures of Ideas,* stated the five qualities of civilization as Adventure, Beauty, Truth, Art and Peace.

All five can be found in the North, where life or death can hang in a turn – in unreadable ice, hidden river currents, blind valleys, and volatile weather patterns. Such natural phenomena, along with spiritual legends that can be neither proved nor disproved, caution people to treat places like the Nahanni with respect.

With modern communications systems and better equipment, and perhaps additional knowledge and respect for the wilderness, the deaths and disappearances have slowed in the Nahanni National Park in the past 30 years.

One of the latest, in June 2006, was the search for a canoeist who had been enjoying a guided paddling trip with his daughter down the swift-flowing waters of the South Nahanni River. When the canoe upset, a guide was able to rescue the daughter, while another guide pulled the man from the water and began to give him artificial respiration from her canoe. However, as her canoe floated into rapids she had to let go of her patient, who was swept downstream. A call for help was sent out over open airwaves, asking any nearby aircraft to help with the search and rescue effort. A deHavilland Twin Otter, that had earlier transported Canada's Governor General Michaelle Jean on a flight over Nahanni National Park, rallied to the cause and soon spotted the body of the 50-year-old man, Michel John Boucher.

Then, on August 5, 2009, the military Rescue Co-ordination Centre (RCC) Trenton reported a helicopter crash west of Nahanni Butte, with three people aboard. In the words of the report:

> *C-GNYT, a Robinson R44 owned by Wild Water Heli-Fishing and operated by Nahanni Outfitters was landing on a small pad adjacent to a steep slope when the helicopter tumbled down the slope and caught fire. The two passengers are deceased, and the pilot suffered serious injuries. The RCMP and multiple local resources responded including several helicopters. Reports indicate that the extraction of the survivor [would] be difficult due to the proximity of the closest landing pad and a steep slope. A SAR C-130 Hercules from Winnipeg was also tasked.*

Fortunately, the pilot, while severely injured, was safely rescued from the crashed helicopter and was flown first to Fort Simpson and then to an Edmonton hospital.

Cause and Effect

Jimmy "Midnight" Anderson for years flew his Piper Super Cub from his home at Pink Mountain, Mile 147 on the Alaska Highway, into the Tuchodi mountain ranges, over the Valley of Lost Planes, and throughout the Nahanni.

He offers practical answers to some of the mysterious disappearances.

> *In the early days, when people went in and didn't reappear, there was no chance of anyone finding them. The territory was unexplored, with 3,000-foot cliffs of old prehistoric rock going down to the water. There are downed planes all over in there. You can't see anybody or anything through the trees and those mountains.*

There's just one way out of the valley – down the South Nahanni River. Those who meet their deaths in the inner reaches of the Nahanni, often by falling through the river ice, are carried downstream to be deposited on beaches and gravel bars, or snagged in low-hanging tree branches. More than one unsuspecting summer visitor has been greeted by such a ghastly sight.

"The Nahanni flushes out all its debris," Midnight Anderson concludes.

Sometimes the bodies are headless.

That too might have a natural explanation. "If someone falls through the ice they're usually well-dressed in winter clothes, with just their faces exposed," reasons Anderson. "Well, as the body washes up on shore the birds peck away, starting from the face and going into the parka hood. Then the larger predators come along and pull everything apart. People in boats discover the headless body and that keeps up the legend. Really, it's just nature."

Gregson's Gold

The Discovery gold mine, located some 55 miles north of Yellowknife, was humming. Beginning with the first gold brick produced here on February 10, 1950, by Consolidated Discovery Yellowknife Mines Ltd., the mine site had grown to include a couple of bunkhouses that accommodated 50 workers, a cookhouse, a recreation hall and office building, several residences and a warehouse. But there was no radio, or even telephone communication. With no road to connect this isolated mine to town, the main means of transportation to Yellowknife was by air.

The workforce of such a place is always comprised of people who've come from elsewhere. One such person employed at Discovery Mine was Tony Gregson, a handsome young Englishman of about 30 years of age. The year was 1954.

At first Gregson worked in the mine's office, but soon found himself in trouble with the boss. After a weekend off in Yellowknife he apparently "forgot" to return to work, and was gone for a week. When he finally showed up, supposedly walking the full distance to get there, his boss decided that Gregson needed an attitude adjustment and assigned him to the worst job in camp – operating the "honey wagon".

Angered and humiliated, Gregson formulated a plan. At that time, mine management was rather cavalier regarding the handling of their product. Every week the mine poured two gold bricks, each weighing 48 to 50 pounds; with gold at that time costing between $32.50 and $35.00 an ounce, the combined value of the two bricks was approximately $55,000. The bricks were sewn into canvas bags and marked with inventory numbers and their destination, in this case the Royal Canadian Mint in Ottawa. The bags containing the gold bricks were then included in the mail sack and loaded with other cargo into the back of bush pilot Max Ward's chartered DHC-3 Otter (CF-GBY).

When the floatplane docked at Yellowknife and passengers had disembarked, the cargo would be unloaded onto the dock. Frenchy's Transport, a small local expediting company, would pick up the mail sack and take it to the post office. If it came in after office hours, the sack would be taken into Frenchy's office, also located on the dock, for storage until the post office opened. Then the bricks would be weighed, registered and shipped by post to Ottawa.

How could anything go wrong? No one could steal from any of these gold mines – Consolidated, Discovery, or Giant – and hope to escape. The Northwest Territories was an isolated place, with just two ways out. From Yellowknife one could fly south via Canadian Pacific Air Lines' scheduled daily flight, or get to the town of Hay River on the south side of Great Slave Lake and from there take the bus. The police could easily detain both these transportation facilities if an alarm was raised about some missing gold.

Gregson was seething as he drove the honey wagon truck around the camp, loading the slop. He had been an office clerk, and now was forced to perform the most stinking job in the place. A few guys smirked when they saw him coming, but most were sympathetic. Gregson was well liked, and the boss was being a jerk. Everyone said so. The punishment did not fit this crime. Who hadn't gone AWOL once in a while from these isolated camp jobs? It would drive anyone daft after a while.

Gregson had seen the gold being poured, and the pure gold bricks snuggled into canvas sacks, to await their ride to Ottawa on the mail plane. Heck, anyone could pick them up! It was just that nobody had thought of it before. Nobody had a reason to think of it before. Now, he did. Those two little bricks would set him up for life. A nice house in the city could be bought for under $10,000, and the fanciest Cadillac for less than $5,000.

He began to plan. The best idea would be to replace the gold with something else. No one ever looked inside the sewn bags until they arrived at the Mint. The only time the bricks would be handled was when they were weighed and registered in the post office in Yellowknife – and even then they were not removed from the bags.

Gregson continued to make daily rounds with the truck, waiting for his days off. At the next opportunity he booked a flight to Yellowknife. En route, he carefully watched the operation: the mail sack was hauled out of the airplane, placed on the dock and picked up by Frenchy's Transport. No one even looked at it, just a heave-ho and it was onto the truck and over to the post office, or Frenchy's office.

While on leave, Gregson contacted a seamstress in Edmonton and gave her an order to sew two canvas bags, of a certain material and dimension. Step One.

Back in Yellowknife, he paid a visit to Mayor J.G. "Jock" McNiven, manager of the Negus Mine (and the first appointed Northern Councillor of the Northwest Territories), and bought 150 pounds of lead "to fix his boat." *

It would not be difficult to melt down the lead – a cast iron pot over an open fire would do – and pour it into bread pans to form bricks to imitate the size and shape of the gold bricks. Gold weighs about five percent more than lead so he'd have to mould the bricks close to the right size without making the weight difference noticeable. Luckily, they were included in a sack with other mail so the pilot or the transport people shouldn't notice the difference when they picked them up.

He carefully wrapped his homemade bricks into the canvas bags. Now, what did they write on them? Oh, yeah. "Royal Canadien Mint, Ottowa, Ont.," he pencilled, not realizing he'd misspelled a couple of words. Then he copied the numbers from the last shipment. By the time that was discovered he'd be long gone.

His escape plan was the next matter. He was an avid canoeist and knew the dangerous and erratic waters of Great Slave Lake. Some of the guys had laughed at his little boathouse down on the lake near the "Con" mine. He had to admit it looked kind of strange, poking up like a telephone booth, but it kept his canoe stored in an upright position, out of the weather, with room inside for his small outboard motor. Going along with their joke, he'd even made a sign, "Tony's Boat House in the North." Some joker had stuck up another sign, "Looks more like a two-story Outhouse in the South." But Gregson would have the last laugh.

Between Yellowknife on the north side of Great Slave Lake and the town of Hay River on its south shore (where the road from the south actually terminated) lay a span of 90 miles of churning water, or a shoreline trip of nearly twice that distance. In preparation for his getaway to coincide with the July 1 holiday, Gregson got into his canoe and then hid it and the outboard motor from view near the float plane dock in Yellowknife, ready for his escape south by water.

He would fly from the mine to Yellowknife with the other camp men and cargo, and switch the sacks en route – lead for gold – on the aircraft. Once in Yellowknife he'd hail a taxi from the dock to town, and when the coast was clear he'd return to the dock, hop into his canoe and be off.

---

* *The Story of Discovery Mine* compiled by the Discovery Women's Institute states that Gregson bought one lead brick for $25.00 from a foundry and a second at a later date for $30.00 from a watchman at a non-operating gold property in Yellowknife.

Right on schedule, Max Ward came in to Discovery Mine with his float-equipped single engine Otter, painted in the new Wardair colours of blue with red and white trim. Gregson was ready with his ticket to Yellowknife. Max Ward loaded the cargo and the mail sack containing the two newly poured gold bricks for the short 55-mile trip while Gregson and two other passengers climbed aboard.

In order to be left alone, Gregson developed a scheme: he'd pretend to be drunk, not an unusual situation for miners ready for a whoop-up in Yellowknife. On the way, Ward had to drop off some groceries and pick up mail at a survey camp, and also stop at Consolidated's woodcutters' camp a few miles to the south where a crew was doing maintenance work on the power line to Discovery. Good. The passengers stayed on the aircraft while Max Ward went out onto the float, handing off the freight and throwing in new mail sacks.

Continuing his drunken act, Gregson leaned over the mail sack that lay beside the pilot's seat, as if losing his balance. In the confusion of righting himself and sorting out items that had fallen, he deftly made the switcheroo – his "mail sack" for the one from the mine containing the gold. No one noticed, but Gregson could barely breathe during the remainder of the short trip.

Yellowknife quickly came into view. Ward taxied to the dock, tied up and began the chore of offloading the cargo, swinging the bags down from the sill of the aircraft onto the dock. Tony noticed that Ward gave an extra heave-ho as he swung off his bag.

"My God, what's in here, gold bricks?" Ward asked jokingly.

Gregson gave a little laugh but said nothing, as Ward hadn't looked directly at him. In fact, he wouldn't know whose bag it was. But close call, that.

With baggage in hand, Gregson scurried off in a taxi. Later, he returned to his secluded canoe and turned the bow out into the dark waters. When he was a short distance from shore, however, the motor quit. What now? He looked around – big waves threatened to upset the canoe no matter which way he turned. Maybe it wasn't such a good plan to brave these waters all the way down to Hay River. He'd once gone on a canoe trip to the Nahanni region with three others fellows, and had been the only one to return. The others had drowned. Awful thing. Perhaps fate had intervened once again to save him from a watery death.

He paddled back to the dock late toward evening, and went into town to make a couple of phone calls. The first call was to Jim McAvoy, who flew a Stinson for his family company, McAvoy Diamond Drilling. "No, sorry," McAvoy replied, "I just can't take you to Hay River right now. Too busy."

Wardair was risky. What if Max Ward had figured out he'd switched the bags? But, there were no others.

Shorty Brown was with John Parker at the Wardair dock watching cargo being unloaded. He was surprised to see Gregson return.

"Shorty, is that plane over there, that Beaver, for rent?"

"Yeah, it probably is," Brown replied. "Go inside and talk to Max Ward."

"I have to get to Hay River, right away," Gregson explained, seeming flustered. "Family emergency. I can't wait for the regular CP Air flight on Monday. I'll pay the full six-seat fare in the Beaver, in cash. Just get me there, first thing Sunday morning. Gregson's the name. Tony Gregson."

Max Ward felt that something was odd. First, it was pretty expensive for a chap from the mine to charter an aircraft. Also, he'd just brought this passenger in on an earlier flight from Discovery Mine and there'd been no mention of an emergency. Maybe he should phone the RCMP, as he usually did to report on unusual movements. He had his finger on the dial when a taxi cab pulled up in front of the office. He'd make the call later. Just take this guy's money and call pilot Henry Hicks to fly him to Hay River in the Beaver.

The trip was uneventful. If the pilot wondered why someone wanted to pay full fare for a six-seat charter, it was none of his business. He was being paid by the flying mile and the extra income would be useful. Hicks' attempt at casual conversation went unanswered. His passenger in fact seemed a bit nervous, but didn't Max say the guy had some kind of family emergency?

On reaching Hay River, Gregson nodded farewell to the pilot, and caught a ride into town. There he purchased a bus ticket for Edmonton, leaving Monday morning.

Before boarding, he wrote a short letter to the manager at Discovery Mine, and dropped it off at the post office. "Give my pittance of a salary to charity. Good-bye. Tony Gregson".

Sitting on that long bus ride he considered his next step. By the time the bus reached Edmonton, the word would have got out about the switch and the police would be waiting for him. When the bus pulled into the town of Peace River, he got off.

After enjoying a meal and a walk around town, he returned to the bus depot and bought a ticket for Prince George. Once there, he purchased a train ticket to the farthest destination he could reach by land on the B.C. coast – Prince Rupert. This would take him far away through miles of forested wilderness of British Columbia. Good!

Four days later, "Anthony Johnson" bought a ferry ticket for Vancouver… and Tony Gregson disappeared.

Down North

When Max Ward placed the mine's mail sack containing the gold bricks on the end of the dock for pickup by Frenchy's Transport, John Parker, a local mining engineer, had been present. Noticing Ward standing around as if waiting for someone, Parker asked him what was up. "Oh, there's the gold bricks from Discovery," Ward said, indicating the bag. "Just keeping an eye on them."

"What do they feel like?" Parker asked. "Are they heavy?" He bent to pick one up. "What is this supposed to weigh?"

"Fifty to seventy pounds each," Ward replied. "They're marked right on the tag."

Parker again hefted each bag. "They feel quite a bit lighter than that."

"Don't be ridiculous," Ward responded.

"Oh well," Parker responded, putting the bags down. "I thought I was a better judge of weight than that."

Nothing more was said. Frenchy's Transport picked up the bags, and Tommy Nuttall put them in the office safe until the post office opened on Monday.

First thing Monday morning the canvas bags from Discovery Mine were taken to the post office to be weighed and registered in preparation for mailing.

The numbers on the bags did not at all match those on the documents. The numbers were those from the *last* shipment! The weights were wrong! And why was the destination spelled incorrectly? They called mining engineer Norman Byrne, brother of Consolidated Discovery Gold Mines president Jerry Byrne, and when he arrived the bags were opened.

"Lead!" cried one man.

"Lead, eh?" echoed a second, as they found themselves staring at the two crudely poured loaves.

They stood staring at the useless cargo in front of them, and then phoned mine manager R.J. Kilgour. He asked them to weigh the bags. When it was discovered that they weighed 48.2 and 43.2 pounds, rather than the recorded 72 and 52 pounds that Gregson had printed on the bags, they knew they had a crime on their hands.*

Max Ward now recalled, with embarrassing clarity, lifting a passenger's bag from the aircraft and jokingly asking if it contained gold bricks. He also recalled his oddly behaved passenger Tony Gregson and the man's desperation to book a charter flight out of Yellowknife. At that moment Max Ward knew he had been party to an incredible scam. Now, thanks to the efficiency

---

* Max Ward states that the bags were marked to weigh 48 pounds, and John Parker said they were too light (*The Max Ward Story,* pp. 130). *The Story of Discovery Mine* states that the bags were found to weigh 48.2 and 43.2 pounds instead of 72 and 52 pounds. It is probable that Gregson did not have the exact loaf mould used for gold casting, and his lead loaves were smaller than the gold ones, since the weight differences cannot be accounted for by the greater density of gold over lead (approximately five percent).

of his own chartered Beaver aircraft that had flown him out of Yellowknife, the crook had a three-day head start.

"He is not someone I am ever likely to forget," Max Ward states in his autobiography, *The Max Ward Story*, "and, while the loss was covered by Discovery's insurance, the shock to my system when I found I had been duped was considerable."

Corporal Bill Campbell of the RCMP was put on the case, with charges laid against Anthony Gregson the following Tuesday and a warrant issued for his arrest. The letter directing his "pittance of a salary" to charity had been received, as well as disclosure that he'd declared the beneficiary of his company life insurance policy as the Vancouver Symphony Orchestra.

The Yellowknife weekly newspaper, *News of the North*, had a field day with the story, a definite Page One item for its July 9, 1954, issue. In the article, titled "$50,000 Worth of Gold Spirited Away", the editor couldn't help add admonishment regarding the lackadaisical manner in which the gold was handled by the mine:

> *Up to the present Yellowknife's great protection has been its isolation, but whoever is responsible for last week's theft of two bricks from Discovery figured out a way to make that isolation work for him.*

Many of the locals approached by editor Ted Horton acknowledged that they too had at times entertained the thought of slipping a couple of ounces of gold bullion into their bags. The editor stated:

> *We have frequently heard even law-abiding, upright and honest citizens of Yellowknife day-dreaming audibly about how just such a theft could be done. Though actually few of the virtuous would have the slightest idea as to what to do with a gold brick, if they had it, it was still regarded as a tempting idea.*

The idea seems to have been that while stealing a gold brick is a crime and therefore an act to be deplored and frowned upon – it's still not a sin.

Max Ward admitted that the affair caused him to become more cautious with his cargo. "Thereafter we took very good care of gold bricks when we had to move them," he said, on the CBC radio documentary *Gold Heist*. "We continued to move them as we did before, but they were under my seat in the aircraft, and if I went out of the aircraft I checked them when I came back. Yet, we weren't as secure as having a Brinks truck or anything like that around."

AWOL

Three years passed. On the streets of Yellowknife and in the bunkhouses at the mines, people continued to theorize on where Gregson might be, how he'd cashed in the gold, and what he'd done with it.

Rumours circulated that Gregson was living the good life, shaving off bits and pieces of gold and selling them as needed. Some heard he'd sold the gold in Vancouver for much below value. Others that he'd bought a boat in the Maritimes and sailed to the Caribbean. He was enjoying some high living in Los Angeles. He blew the money on racetrack betting. He had gone to Cuba where his fortune had been stolen from him.

One rumour guaranteed continued interest in the story – some of the gold had never left the North! One brick was buried near Yellowknife, in Old Town, where the airplanes docked and where Gregson's canoe had been launched and then abandoned. It made sense. If he was caught, he'd lose only half his booty. No one would ever know where he'd cached the remainder.

Then, on June 27, 1957, the speculations came to a dead stop when *News of the North* announced, "Tony Gregson Without Bricks Picked Up in Australian Gaol." He had been caught on a relatively minor offence for stowing away on a ship, *S.S. Cornwall.* While serving a short stint in a Sydney jail, the outstanding Canadian warrant had surfaced.

Gregson was brought back to Yellowknife to answer the theft charges, and on August 19, 1957, he appeared before Justice J.H. Sissons. In the August 22 issue of *News of the North*, the public was treated to a rather short "facts only" item:

Gregson's plea is guilty of theft

Tony Gregson has stopped running. No longer need he look over his shoulder nor writhe when he sees a policeman.

He pleaded guilty here on Monday to the theft of gold bricks from Discovery some three years ago and he was sentenced to two-and-a-half years in jail, which he will serve in Prince Albert.

Offence is said to have been committed by him by substituting lead blocks he had fashioned himself and wrapped to resemble gold in transit.

He appeared for his preliminary hearing before Magistrate L.H. Phinney at nine o'clock on Monday morning, was remanded to a higher court, and appeared before Mr. Justice Sissons at 10 and was sentenced by 10:30.

Yellowknife citizens had anticipated a long drawn out legal battle and many felt Gregson would have gotten away.

> Terry Nugent of Edmonton appeared for the Crown and D.J. Hagel on behalf of Gregson.
>
> It is not thought he has disclosed what disposition was made of the gold.

The Northern justice impressed the local people. The criminal neither got off scot-free nor was he overly punished. While no decent Northerner would actually rob a fellow miner's gold, to steal from an insured company was another matter entirely. This was "high-grading" of the highest order. If Gregson's picture were to appear in the newspaper, everyone was sure he'd have a grin on his face.

But not everyone was laughing about the story. "If he ever turns up in Yellowknife, I will recognize him," Max Ward stated grimly.

But Gregson *did* return after he'd served his sentence, sneaking in under Ward's wing, perhaps even in one of his aircraft. Bing Rivett, a former RCMP officer who was now a personnel manager for Giant Mines, states that Gregson not only returned to the scene of the crime, but also came and asked him for a job. The answer, he says, was an emphatic "No!"

Might Gregson have returned to search for the brick he left behind? Or was he just seeking a new start with a well-paying job? If Gregson was so broke that he had to stow away on a boat, folks reasoned that he must have gone through all the gold.

Yellowknifers later heard that Gregson had found work in mines, from Manitoba to Quebec, but would suddenly quit and leave if someone who knew him from Yellowknife appeared. And so he became a wanderer, chased from job to job, paying forever for his daring deed. But, it might have all been worth it: that sweet revenge against a tough boss; the making of a Northern legend, although it set his life on a never-ending flight path.

Gregson died in September 2001 of heart disease. And, while the details may have disappeared with him, the tale did not. After all, this was the North, where such stories live forever, to be told and retold – perhaps embellished with each telling but with a nugget of truth still glinting at their centre.

The Unfair Weather of Mount Fairweather

On July 20, 1951, Flight 3505 (a Canadian Pacific Air Lines DC-4, registration CF-CPC) left Vancouver International Airport at 18:53 en route to Tokyo, Japan. The passenger list for CF-CPC that evening comprised 29 American servicemen and two Canadian sailors.

Canadian Pacific Air Lines (CPAL) had only recently expanded to overseas routes, after earning a coveted operating permit to provide scheduled air service from Vancouver to the Orient. The outbreak of the Korean conflict in June 1950 brought CPAL a much-needed contract from the United Nations to transport American and Canadian troops and supplies to Tokyo. However, the 6,500-mile flights northwest over the north Pacific would prove challenging for CPAL and early air traffic in general.

There were two possible routes a flight could take for the first seven-hour leg of the trip, between Vancouver and the U.S. Air Force base near Anchorage, Alaska. These major "highways of the sky" followed radio range navigation systems designated by U.S./Canadian aviation authorities.

The "commercial" route, called *Amber 1,* followed an airway that doglegged along the coast, with radio range legs that ran parallel to the mountains. In the 1950s, flying along this route required the precise aural monitoring of several consecutive stations. From Gustavus, Alaska to Anchorage, the mountains were as close as 25 miles to the airway, so it was normal practice for aircraft to fly on the edge of the airway farthest away from the mountains. Aircraft would not have to drift far off course to crash into them.

The other choice, named the "military" route, the one used most often by the RCAF because it flew farther from the mountains than Amber 1, followed the same radio range system to Sandspit, on the Queen Charlotte Islands. It then went directly by dead reckoning over the open ocean to the range station at Middleton Island just short of Anchorage. The military route required more flight planning but allowed a greater margin for error and avoided unfavourable weather, specifically the Pacific lows that moved up the coast. For these reasons, most CPAL flights chose this route.

Storm Warnings

Prior to departure, Captain Victor Fox, First Officer Bruce Thompson, and Navigator Ernest Kraucher were given a meteorological briefing and informed that a weak frontal system was currently forming in the Gulf of Alaska. The front would be 150 to 200 miles off the Gulf Coast, between Sitka and Yakutat, at the time their flight would be passing through there. Although light icing conditions might be expected, this was not unusual, and no warning was received regarding severe icing or turbulence on either route.

First Officer Bruce Thompson, a former aircraft captain in New Zealand who was known to be a conscientious and capable pilot, commented that the weather was generally better on the military route: "It's over the open sea and away from land," he commented. "Along land, near the mountains, the clouds tend to pile up and are thicker with more icing and turbulence." Since the flying would be shared by the captain and first officer for various legs of the journey, his input was warranted.

However, Captain Fox disagreed, stating that "navigation was easier along the inside, commercial route." It is possible that the captain, perhaps lulled by an anticipated routine weather situation, might have selected the Amber 1 route because of the easier navigation: one pilot could easily operate the aircraft. The DC-4 was equipped with an automatic pilot system which did the flying while one pilot monitored the flight status and the others could rest on the long flight. The workload in the cockpit was minimal if all was going well.

The captain's choice prevailed and the first officer apparently shrugged and offered no further disagreement, although observers sensed his discomfort. Just prior to departure, Thompson passed by the dispatch office and muttered, "I wish I wasn't going on this trip." The dispatcher nodded sympathetically, but he was used to crews making similar comments. Although Thompson's statement seemed to convey a deeper reason, nothing more was said.

The flight took off, heading north into darkening skies.

In line with normal operating procedures, the radio operator reported in as the flight progressed along Amber 1, passing Comox, Port Hardy, Sandspit, and the Dixon intersection near Prince Rupert. However, the next position report, due to be sent from Sitka to the control station at Yakutat, arrived 23 minutes late, at 23:00 hours. Whether they had encountered a strong headwind prior to Sitka or some other problem, this detail seemed irrelevant at the time, especially when the aircraft indicated no difficulty when it passed over the next radio reporting point after Sitka, the Cape Spencer intersection. In fact, the voice message specified, "All's well." The aircraft was on course, now within three minutes of the flight plan, possibly indicating a shift in wind direction from a headwind to a tailwind, and still at the assigned flight altitude of 9,000 feet.

The minimum altitude over this mountainous coastal route was 7,000 feet, although it was recommended that this be increased during turbulent weather conditions. If the aircraft went too much higher, however, it might encounter the stronger headwinds that had been forecast. And because the DC-4 was not pressurized, flying at higher altitudes would have been uncomfortable for passengers and crew.

The weather conditions at both Anchorage and Fairbanks, the alternative landing field, were good, and were expected to remain so for the planned arrival of Flight 3505. The next ETA [Estimated Time of Arrival] radio check was scheduled for midnight at Yakutat. It was never received. The aircraft had disappeared for no apparent reason two hours short of its first stop.

Strange phenomena

Fourteen hours after the last communication was received, a faint radio signal lasting over three hours was picked up at Anchorage, at Cranbrook, B.C., and by ships at sea. Although as usual the RCAF had all coastal radio stations set to the international emergency frequency, none could get a bearing on the steady hum, which revealed no Morse code signals. Finally the Coast Guard determined that this signal could not be from the missing airplane.

If the aircraft wasn't in the water, what had happened to it?

The predicted shifts in wind direction as the aircraft entered a weather front between Sitka and Yakutat might have lowered the freezing level, making the aircraft more susceptible to icing conditions. There was also some indication that the air was unstable, with heavy cumulus clouds rising above 9,000 feet, making for low-visibility conditions, and potentially heavy turbulence, just as First Officer Thompson had warned.

A southbound Northwest Airlines aircraft flying at 10,000 feet at about the same time on the same route reported lightning, reduced visibility, and light icing. Stronger winds in the area could have given the northbound DC-4 a greater ground speed and affected its course.

The southbound plane also warned of a variance on the radio range beam that the CPAL flight would have been following when it disappeared. The southbound leg of the Yakutat radio range signal was apparently shifting — something that might occur if there were thunderstorms in the area. Although the aircraft was on a constant heading, the beam would shift or "swing", splitting into several similar beams, some straying outside the 10-mile-wide normal airway. In this situation, a pilot or navigator might have to go "course hunting" by continuously changing headings from the original bearing. If a pilot/navigator maintained the wrong heading, the aircraft could easily fly five to 10 miles off the airway.

Immediately to the east of the Cape Spencer radio range intersection, within a distance of 30 miles, is a series of nine mountain peaks that all extend above the 9,000-foot level of CF-CPC's cruising altitude. From the centre of the Amber 1 airway, the 15,300 foot peak of Mount Fairweather was just 20 miles away, sitting 10 miles inland to the north of Cape Spencer. There was little separation between the adjacent peaks, which made the mountainous wall along the coast a very nasty place for an aircraft flying at 9,000 feet in poor weather.

Freezing conditions that night at the 9,000-foot level near Cape Spencer would not be unusual, for "Fairweather" was definitely a misnomer for the mountain and the area. Another pilot, flying northward along the B.C. coast, recalls his aircraft becoming covered with a heavy layer of ice within seconds, causing it to stall and begin a rapid descent. With engines at full power, he achieved lateral control but continued to descend. He contemplated whether he'd be smashing into the mountains or diving into the sea. After descending 6,000 feet, however, he entered warmer air and rain. Fortunately, the ice melted and at 3,000 feet he was able to level off. Other stories emerged about the oddities of flying in that area, but none offered clues to the whereabouts of CF-CPC.

The Search

Despite the extremely unco-operative coastal weather, an active search for the missing DC-4 commenced at dawn on July 21, involving personnel and aircraft from Canadian Pacific Air Lines, the Royal Canadian Air Force, the United States Air Force, and the United States Coast Guard. On the first search day, a storm centred in the Gulf of Alaska brought drizzling rain and a 500- to 1,000-foot ceiling.

On July 25, searchers happened to spot the wreckage of a USAF C-54 that had disappeared the year before, in July 1950, en route from the USAF base near Seattle (McChord) to Anchorage (Elmendorf). Although that wreckage had been located earlier, all six crewmembers had perished and their bodies were unrecoverable from the 9,000-foot-level of Mount LaPerouse, 30 miles from Cape Spencer. Also on the minds of the searchers was another USAF C-54 that had gone missing six months earlier, in January 1950, while en route from Anchorage to Great Falls, Montana, carrying a crew of eight with 36 passengers. It was never found. That aircraft had radioed from the same radio intersection while on its flight south, as had the CPAL DC-4 while heading north.

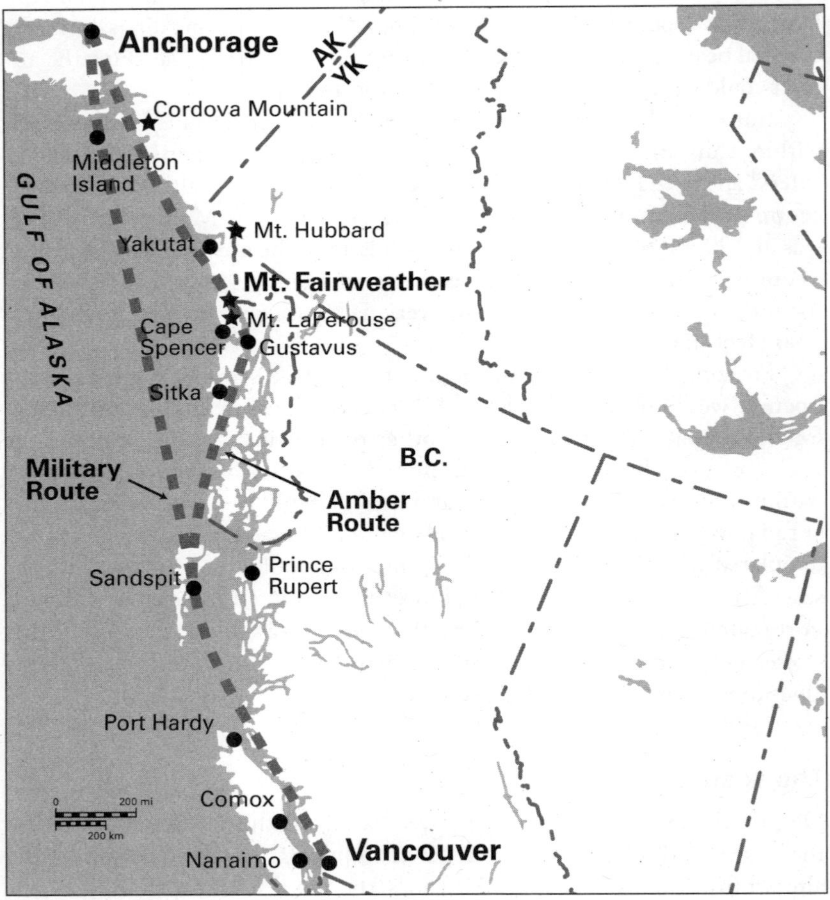

Flight 3505 Goes Missing

Searchers for CF-CPC continued to look for anything out of the ordinary – movement, colour, or displaced patterns in rock or foliage. Although glaciologists from the Juneau ice field research team had determined there was no possibility of the large aircraft being buried under a snow slide or wedged down into a glacier, the spotters strained to identify any possible clues through the fog, snow, sleet, and miles upon miles of rippling glacial ice. Flares sending up heavy black smoke were sighted in the mountains northeast of Prince Rupert, but search aircraft could find neither the source of the flares nor any sign of the aircraft they were so desperately seeking.

The search extended over sea as well as land. The missing DC-4 carried three rafts (two large, one small) which contained fishing gear and emergency rations, as well as an emergency radio transmitter. One of the life rafts on board the aircraft was equipped with a "Gibson Girl", an emergency radio powered by cranking a handle and holding the antenna in the air. In the automatic mode it would send an SOS message in Morse code for as long as someone cranked; in the manual mode a detailed message could be sent. Therefore, a message could have been transmitted via the Gibson Girl if the aircraft had gone into the sea and some personnel had managed to clamber into the raft to crank the handle.

All oddities were checked out: a yellow item floating in the sea 30 miles southeast of Yakutat; a nautical chart catalogue and plywood signal panel near Knight Island in Prince William Sound; a case of rations found by a fisherman near Cordova. None was from the missing aircraft. Some residents of Baranof Island south of Cape Spencer (on which Sitka is located) reported hearing a low flying aircraft the night of the DC-4's disappearance. The land and fjords were searched once again. A fisherman reported seeing some lights in the mountains northeast of Petersburg but still nothing.

Then, in the same search area, a Norseman went missing, with three on board. This aircraft had been assigned to pick up three ice-field scientists at the 12,000-foot peak of Mount Hubbard. Searchers were then diverted to rescue these three stranded scientists. The Norseman was never found.

On August 8, a human leg and shreds of green trousers were found on a beach near Yakutat. Examination showed the leg had been in the water for two weeks, which would correspond to the time the DC-4 went missing. But the leg and trouser fragments proved to be unrelated either to each other or to CF-CPC.

The searches for other missing aircraft continued to interrupt the hunt for CF-CPC. While two U.S. Navy aircraft that had been reported missing on August 13 and 31 and a civilian Cessna missing on August 15 were quickly tracked down, a C-47 lost on September 1 required five days of searching before it was found.

The formal search for CF-CPC was called off on August 31, but continued on an *ad hoc* basis until the end of October. The only conclusion that searchers could make was that the aircraft had crashed on a rocky peak – probably between Fairweather and Cordova mountains.

For Grant McConachie, president of Canadian Pacific Air Lines, this would be the fourth fatal crash of a CPAL aircraft since he had become head of the company. The event was devastating to him as well as to family and friends of the passengers and the crew of CPAL's DC-4, all of whom were anxiously awaiting any news that might bring hope, or at least closure. Telegrams came in to the company as well, expressing sorrow at the incident, and the hope that it would not set back CPAL's attempts to provide global airline service. Lt. Gen. Lawrence S. Kuter, a senior USAF officer appreciative of the Korean airlift operation and CPAL's involvement, wrote:

> *Please accept my deepest regrets of all personnel of the MATS [Military Air Transport Service] in connection with the loss of a DC-4 en route to the Far East in support of the United Nations in Korea. The record of CPAL ... is one which we all admire.*

What happened?

Accident investigators found no obvious signs to indicate what might have put crew and passengers of CF-CPC at risk that fateful night of July 20, 1951. The aircraft had been purchased the year before and was powered by Pratt & Whitney R-2000 engines, totally reliable systems, and sufficient fuel had been on board. The formal investigation thus hinged on the three leading potential causes: the weather encountered, the vagaries of the chosen route, and the judgement of the flight deck crew.

Analysts pondered all probabilities, and concluded that while the weather report issued earlier had given no reason for extra caution, the weather did in fact turn out worse than the forecast.

The crew was rated "competent" to fly the challenging radio range route of Amber 1, but circumstances may have conspired to confound their judgement: they were 20-25 miles from the mountains with poor visibility and the radio data were difficult to interpret. Pilots learned to listen for "beam passage" – that is, the point where the instruments emitted a solid or steady tone indicating that the aircraft was passing through the "beam" separating one coded section from another. But the procedure was tricky, as described by Don Connolly, a former navigator with 426 Transport Squadron, RCAF, who flew the Korean Airlift:

> *At one point on the flight I could hear the signal swell and blend as we approached one beam that pointed almost straight out to sea, intercepting our flight path at right angles. But as the strength of the signal mounted higher and higher and the solid tone persisted, my initial bemusement turned to apprehension. We seemed to be sliding along the beam, virtually drifting sideways towards the coast. It defied reason! Had a 40 knot wind suddenly more than tripled in speed? Finally, as the grow-*

ing signal volume began to assume the role of an alarm, I ordered a 30-degree turn to seaward.

By now my ears were straining to detect a change in signal tone and strength, and my eyes were stuck to the unwavering needle of the radio compass which still seemed to indicate a remarkable 50 degrees or more of drift. I called for another radical alteration of heading – 30 degrees more – and finally we started to break through the beam and away from inhalation by the mountains. An hour later we made a belated landfall at the Queen Charlottes, approaching from an angle 70 degrees askew from our planned track. I bought the beer that night – gladly.

Crewmembers on CF-CPC could have quickly become overwhelmed by conflicting information, especially if they were struggling to confirm their position with alternative methods such as LORAN (a long range navigation system using ground-based stations rather than satellites; this was a predecessor of Global Positioning Systems), which could also be unreliable if there was interference from mountains, clouds, and fog. Quick decisions often had to be made based on information from secondary navigation systems.

The question of why the flight crew was late reporting the Sitka radio range passage remains unanswered. Could it have been because of poor radio transmission? Or difficulty in flying the radio range, or just lack of attention? Or might it have been simply due to the late passage of Sitka because of stronger-than-forecast headwinds in the area? When the range was flight-checked three days later, the bearings of all four legs of the route were serviceable.

Radio range navigation was discontinued several decades ago as it was not totally accurate and was too sensitive to weather – it was a bit of an art, that some pilots were better at than others – and was replaced by Very High Frequency Omni Ranges (VOR) and by an Inertial Navigation System (INS). While radio range and VOR both relied on transmission from ground stations to the aircraft, and equipment both on the ground and in the aircraft, INS did not require a ground station; therefore the radio range stations were no long required. Today, modern commercial aircraft are supplementing their ground navigational systems by the use of a Global Positioning System (GPS), which is based on satellite information.

Wayward Winds?

With no airplane, no bodies and no witnesses, the whereabouts of CF-CPC still awaits discovery. Mount Fairweather – or one of its neighbouring peaks – may be holding tight to this secret, but after 16 other airplanes vanished in

B.C. during the years 1951-1960, other theories for the strange disappearances were put forward.

Investigators wonder if CF-CPC and some of the other missing aircraft drifted off course, and fell victim to the "Venturi Effect", now known to occur along the B.C./Alaska coastline. When a constant volume of wind is forced to pass through a narrowing opening such as between two mountain peaks, the wind speeds up and becomes stronger until it can literally drive an aircraft into a mountainside.

At the time, pilots and navigators had little knowledge about these strange high-speed, low-level and localized winds that could rapidly force an aircraft off course, possibly causing it to fly into the vicinity of the coastal mountains. Navigators at the time were only aware that occasionally something odd would occur and as they shared their bizarre experiences, a healthy awareness grew among them – making the military route more popular than Amber 1. In fact, the risks of flying this route were very, very real, for the crew and passengers of CF-CPC, the Norseman, the C-54s, and others who still lie undiscovered among the peaks of this rugged, windswept, and mountainous coastline. The seven crew members and the two Canadian sailors lost on the flight are part of a group of 29 Canadians associated with the Korean War who have no known graves. Their only memorial is the commissioned artwork of CF-CPC, "Toward Their Final Destination", by B.C. artist Dan Ryan, which can be seen at the South Terminal of Vancouver International Airport and which also appears on the front cover of this book.

The Book of Broken Dreams

George Maude of Sidney, B.C., has spent his life involved in the aviation scene. He is the proud owner of a vintage Curtiss P-40E Kittyhawk aircraft, and is a fountain of knowledge regarding the adventures – and misadventures – of his fellow aviators. George maintains a history collection that fills the bookcases lining the walls of his library. Included is a unique book he has compiled himself, bound in wooden covers and too heavy to lift, that could be titled "Death and Disasters". Filled with items pasted onto thick grey pages, the book is bound in wooden covers, measures 27 inches long by 20 inches wide by 5 inches thick, and weighs 42 pounds. In it are clippings and reports on extraordinary incidents related to aviation that have occurred mostly in his area and home province, but also throughout the country.

Dying for Love

One of the oddest stories pasted into the book begins, innocently enough, with the headline "Old planes await auction by public trustee of Ontario". The newspaper article tells of Ernie Simmons's dedication to old warplanes and other machinery no one else seemed to want that he collected on his farm near Tillsonburg, Ontario. The sad twist of this story/obituary in George's book of broken dreams: the poor man may have died for them.

The Vancouver *Province* on August 29, 1970, retold his story, which George Maude carefully pasted into his book, along with postscript headlines from September 4 that noted hundreds of people showing up "to bid on military junk."

Ernie had died earlier that year and, rather than print a straight obituary, the newspapers related the man's obsession with a collection that included an almost unbelievable number of vehicles, aircraft and aircraft parts, weapons, spare parts, and other mechanical artefacts.*

Simmons's love for his collection became his nemesis, and he literally worried himself to death about what might happen to his precious artefacts after he died. On one of many scraps of paper found in his ramshackle house, he'd written, *There are probably more injustices and unfairness in the wrong ones getting a share of a dead man's property than anything else in the world.*

His fears were realized when, in December 1969, six men had appeared at his door, shot him six times, and left him for dead in the snow, robbing him only of the $175 cash he had on his person. When the severely injured man was found and transported to hospital, he quickly signed himself out to return home to protect his stuff; shortly thereafter, in January 1970, he was found dead in the bedroom where he'd been born 57 years earlier. The farm and everything on it was to be sold, and so, stated the newspaper reporter, "his junk will be merely a memory."

While sitting on the floor in George Maude's library, turning the heavy pages of the Death and Disasters book, one feels sympathy for Ernie Simmons, whose devotion brought him misfortune and death. Such a narrative, as well as others in this book, affirms how, in the blink of an eye, unexpected or even catastrophic setbacks can happen. Sometimes care and knowledge are just not enough to overcome bad people, bad weather, mechanical failures – or fate.

Whether the Weather

When a Britten-Norman BN-2A-21 Islander aircraft (C-GIPF) owned by Central Mountain Air Services of Smithers, B.C. went missing on September 2, 1983 on a non-stop flight from Campbell River to Smithers, a search was started immediately. The pilot, Melis (Mel) Melissen, a former base manager/agent in Smithers for Pacific Western Airlines, had loads of experience and his passengers, a party of eight German hunters, were said to have loads of money. So why had these men chartered a rather small aircraft, loaded it to capacity (or more), and then flown in questionable weather conditions?

---

* Included in the collection were eight Fairey Swordfish torpedo bombers as well as 36 other aircraft; 107 motorcycles; 15 army vehicles, 15 Bren gun carriers; 57 rifles, shotguns and handguns; over 125 cars and trucks dating as far back as 1928; 30 aircraft engines; innumerable propellers, tractors, and steam engines; and countless parts, appliances, pieces of farm machinery, and other bits and pieces.

Bill Cove is a friend of George Maude, and a West Coast pilot who studied the case. His son, Dan, had fuelled the aircraft in Campbell River, and was the last person to talk to the pilot before the Islander disappeared.

That day Bill Cove had followed the same route, coming in from Powell River to Egg Island, and landing at Campbell River at 19:35 hours in a Cessna 206 (C-GCRR), where he informed the station "The weather is super-bad." He'd heard and talked to six other pilots in the Johnston Strait area who couldn't make it around Cape Caution and had turned back around Egg Island. "They couldn't get by and they've gone back to Port Hardy," he later reported. "It didn't look good in Smith Sound either, with the clouds right down to the deck."

Cove had come in "between layers", and then it was solid from Knight Inlet north and west, where the upper and lower layers appeared to merge. From Minstrel Island northward it was solid cloud. Cove added a further warning for the "six that were down." He'd encountered problems with the opaque, granular rime ice that sometimes forms on the leading edges of aircraft passing through moisture-laden air.

Descending near Egg Island, Cove was in cloud until 400 feet above sea level. All routes to the interior looked questionable as well, with the front of fog racing eastward. "It was the fastest front I have seen in 30 years of flying," Cove said, "much faster than forecast. The ceilings really collapsed. It was a different day than that which was forecast."

Finally, the search for the Islander was concluded with nothing found, and a major lawsuit was filed against Central Mountain Air and pilot Mel Melissen by the families of the passengers. The courts listened carefully to the taped magnetic log at Campbell River Flight Service Station that included both Cove's warnings and the station's attempts to contact the Islander: "… altimeter three zero one zero with Reported Conflicting Traffic … Islander last reported over the town for [runway] eleven."

Bill Cove himself had flown 1,500 hours, and over the same route, in the type of Britten-Norman Islander that had disappeared in solid fog. In addition he had, at that time, 20,000 hours as a pilot, and held an Airline Transport Pilot rating and Instrument Flight Rules [IFR] endorsement for fixed wing aircraft. With these credentials, and knowledge of both the aircraft and the territory, he was called to make a statement in the Supreme Court of British Columbia action that followed the unsuccessful search.

In his view, Melissen's aircraft, loaded as it was, could not hope to fly higher than 12,000 feet above sea level [ASL]. To complicate matters further, the minimum altitude for that route exceeded the ceiling of this model of aircraft flying at or near its gross weight. Climbing any higher on IFR would bring icing. The aircraft was not equipped with de-icing equipment. Flying VFR [Visual Flight Rules] in that region on that day was also out of the question with the abysmal weather forecast.

At 17:21:17, C-GIPF had called in: "Campbell River – India Papa Fox – we're taxiing for takeoff two nine northwest bound." Campbell River Radio responded, "India Papa Foxtrot – Campbell River – Roger – wind zero seven zero at two – altimeter three zero one zero – no reported conflicting traffic." "Papa Fox checks." At 1731:31 C-GIPF reported "Papa Fox is clear [of the] zone," and they were gone.

The courts listened carefully to the taped magnetic log at Campbell River Flight Service Station that included Cove's weather warnings. At 19:34:58, Bill Cove called in, "Golf Charlie Romeo Romeo … there's about five or six turned back around Egg Island there – they couldn't get by and they've gone back to Hardy … it didn't look good in Smith and Rivers there, it's all right down, the bays in [indiscernible] pocket 502 miles, and that kind of stuff. … We're between layers actually and it's solid from about Knight Inlet north and west there."

Bill Cove, however, wasn't so quick to criticize the pilot for taking that route. "With respect to mountain flying and weather, no two days are ever alike," he stated to the courts. "The terrible thing about mountain flying is that you have to go and have a look. *The guy who's there knows best.*"

Melissen had been in Campbell River a number of times with that aircraft, so he'd been over the route before. In Cove's view, there was nothing wrong with Melissen going up toward the head of Knight Inlet to "have a look at it." He was the only person who had experience *that day* with the weather on *that route* – north of Knight Inlet, which Cove considers to be the most sensible, the lowest and the preferred route – to get to Smithers in the northern B.C. interior ("better than from Bute Inlet and via Hamathko, unless you're going to Williams Lake, in which case the route by way of Bute Inlet is more direct"). But many factors can come into play to change a positive to a negative: the power and performance of the aircraft, turbulence encountered, and the visibility factor. All turned sour on this fateful day.

After eight hours of deliberation the jury found in favour of the plaintiffs – relatives of the deceased passengers – and concluded that pilot Melissen had been negligent. A direct flight from Campbell River to Smithers under instrument flight rules was not possible on September 2, 1983 because the minimum altitude for that route exceeded the ceiling of this type of aircraft when at or near its gross weight, and because of the forecast and actual effect of airframe icing.

Nobody really won that day in court, for the aircraft, its pilot and passengers have never been found.

The Vicious Vortex

George Maude's book also contains clippings on catastrophes that resulted from bizarre combinations of factors that no one could ever predict, or prevent.

On April 22, 1968, at 17:18 hours, pilot Johnny Boak, carrying three passengers, was bringing in a Piper PA23 Aztec (CF-OXR) on wheels to Vancouver International Airport. He was advised that a Boeing 707 was about four miles distant on a practice front course approach ahead of the Aztec and would be making a low approach on Runway 08. "Roger," Boak replied.

While the Boeing and the Aztec were in the final approach area, another aircraft in the circuit was informed about the other two. Its pilot reported at 17:23 hours that he could see the Boeing but not the Aztec, and three minutes later he saw a splash in the water off the end of Runway 08. All that he could see of the Boeing, which was now a half-mile west of the runway, was a typical black jet smoke trail from its engines that still lay over the area of the splash.

Then, debris appeared in the water.

Wingtip vortices from the wake of the Boeing, produced by the heavily loaded aircraft flying at low airspeed, had twisted the Aztec out of the sky while on its final approach to Runway 08. This double vortex wake had settled toward the ground and spread out laterally. In calm conditions this would take about three minutes. This time, however, because of the crosswind, the right wing vortex remained in the approach path of Runway 08. The lighter aircraft, which was rolling to the left at the time of impact while the flight controls vainly called for corrective action to the right, went out of control at an altitude too low for the pilot to effect recovery.

Today, it is a common procedure for Air Traffic Control to caution light aircraft about turbulence from departing or arriving "heavies".

But on April 22, 1968, a fatal combination of timing and unpredictable forces brought the lives of four people to an end, the victims of a vicious vortex.

The Loggers' Discovery

Joe Van Es, a logger from Stillwater, B.C., often flew his Stinson 108-3 over to Gilford Island at the mouth of Knight Inlet, and was one of only a few pilots to land on wheels on a little logging road slashed through the trees on this island. While there in 1993-94 he was told by timber-cruisers that back in the forested part of the island lay the remains of an old aircraft partially covered with gravel and overburden. When the loggers went to check it out, they found the skeleton of the pilot still sitting in the cockpit.

Investigators traced the aircraft's origins through the American registration still visible on the vertical fin, and thereby unravelled the riddle of the Fairchild 24W46. It told a sad tale of a pilot with too little experience who'd attempted to fly his tough little airplane into the changeable weather conditions that prevail on B.C.'s wild coastline.

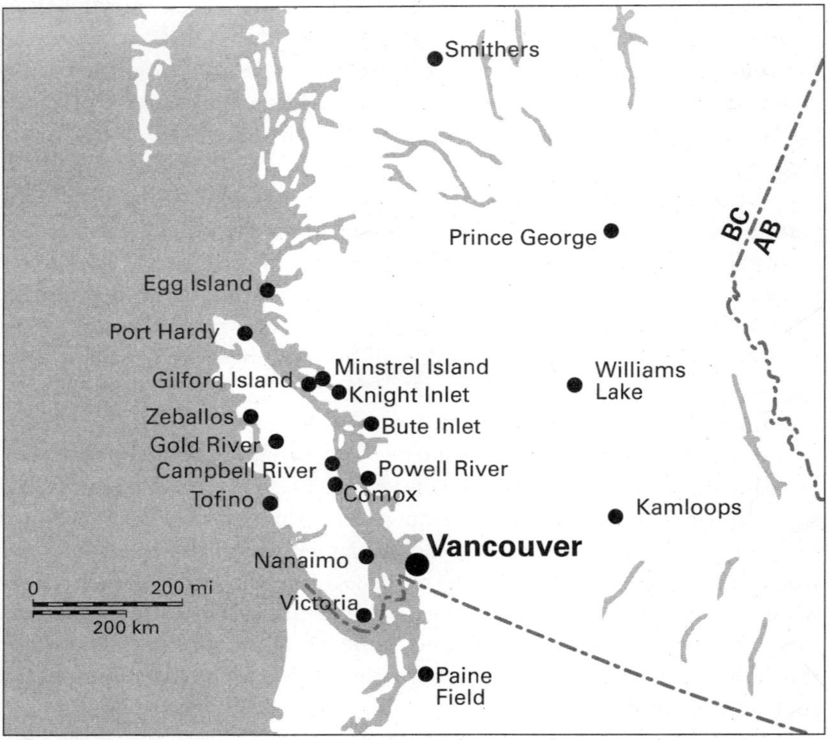

Episodes from "The Book of Broken Dreams"

The Fairchild's fateful trip had begun at Paine Field in Washington State on June 24, 1959. After filing a flight plan to Alaska via Nanaimo, Comox and Annette Island, the aircraft left Vancouver at 10:45 am.

The pilot, a logger with just 60 hours flying time, had not sat at the controls of this aircraft for three years. No problem, with nice clear skies between Vancouver and Comox. But the ceiling dropped to 400 feet with rain and fog, and visibility was reduced to one mile. He was on wheels so he couldn't land on water; he couldn't see to follow the shoreline to Campbell River; with dense clouds extending right to ground level he had no hope of finding Port Hardy to land and refuel as planned.

Then, directly in front of his windscreen, the dark and deadly images of trees would have loomed through the fog. The Fairchild's wooden wing spars hit and splintered, the 165 hp radial engine ripped off to make its own journey, and the red and white airplane plunged into a rocky creek bed.

The flight terminated on Gilford Island, just 45 miles from Port Hardy.

One can only imagine the shock of family and friends when, just before Christmas 1993, they were informed that the pilot and his airplane had been found 35 years after leaving on their ill-fated flight to Alaska.

The 37.5-Mile Flight

On August 4, 1951, a float-equipped Noorduyn Norseman Mark VI (CF-GRQ) piloted by 30-year-old Edward (Ted) Calvin Williams disappeared while on a 37.5-nautical-mile VFR flight from Gold River at the head of Muchalat Inlet to the village of Tofino, on Vancouver Island.

While no flight plan had been registered for the single-engine Norseman owned and operated by Queen Charlotte Airlines Limited (QCA), this was not unusual. Previously unscheduled pick-ups at various logging camps on the west coast of Vancouver Island were common occurrences, and the weather could dictate instant schedule changes.

Ted Williams had been employed with QCA for five months, since March 1951, when he started flying in and out of Zeballos, a Vancouver Island logging town. Williams was properly licensed, adequately experienced, in good health, and was rested after a six-day holiday.

The passengers, most of whom were employed with Tahsis Lumber Company, were familiar with the country they were flying over. Of the six, five were from Vancouver: Larry Larson, Arthur Costley, Charles Warne, Reginald L. Brown and Jack Palmer. The sixth, Ken Mugridge, had lived in Moncton, New Brunswick, until moving to Vancouver Island in 1947 to take a job with the lumber company.

The morning of August 4, pilot Ted Williams and his wife Doris could barely see the hilltops near their home in Zeballos for fog and drizzling rain. "The weather doesn't seem all *that* bad," Ted said as he left the house at 8:45, "but I'll check it out."

Perhaps Williams was more concerned than he let on, for he left behind his zippered briefcase containing maps and his logbooks. But with his pilot's licence tucked inside his coat pocket, and a clipboard with a map pasted on the back outlining the area between Zeballos and Tofino, he'd be fine.

The Norseman's tanks were filled with 192 Imperial gallons of Standard Chevron gasoline, the passengers and their bags were loaded, and they were ready to go. The sea was calm, with winds between five and 10 mph. Just before takeoff the rain let up and the sun started to show, giving a ceiling of 3,000 feet with broken cloud.

The Norseman left Zeballos at 9:00 am for pick-ups at Tahsis, Chamiss Bay, Muchalat Inlet and Tofino. Williams reported in from Tahsis at 10:15 to the QCA agent, William Lloyd Crabbe. When they landed at Muchalat they stayed for an hour, and then departed for Tofino between 11:30 and 12:00. Tofino weather reported a 200-foot overcast, two-mile visibility and light rain – not bad for those familiar with coastal flying. The last positive check occurred as the aircraft passed Danzig Mines, westbound.

Crabbe called Williams several times after that, but received no response. Again not surprising. Radio communication between ground stations and aircraft in this area was chronically poor.

When CF-GRQ was reported overdue at 12:30, Tofino base called Vancouver; the dispatcher phoned Muchalat and notified management.

A search started immediately with the formidable task of spotting a flash of metal, colour, smoke, anything, amidst fog and low stratus cloud along the sharply indented coastline and rocky, precipitous, timber-covered terrain. They diligently carried on until August 21, when the formal search was suspended. Complicating the matter, the Search and Rescue unit was short-handed because of an intensive search for a CPAL DC-4 airliner that had disappeared two weeks earlier.*

So who were the people who disappeared so abruptly and mysteriously that fateful day? And what happened to those they left behind?

Pilot Ted Williams was married with three children. In his RCAF days he had flown Tiger Moths, Ansons, Harvards and Bolingbrokes; and in his civilian life was familiar with Stinsons, Seabees and Norsemen. He'd gained his Senior Commercial pilot's licence in February 1950, and was qualified to fly land-based aircraft and seaplanes up to 12,500 pounds.

Of the passengers, Arthur Costley lived in Vancouver with his wife, Ada. Five of the men were single: Jack Palmer; Larry Larson; Charles Warne; and Reginald Brown, who was engaged to be married. Howard Kenneth Mugridge, the man from New Brunswick, was divorced but left behind four children and a Canadian heritage that had begun in Prince Edward Island in 1818.

Telegrams were sent to each household, with wording similar to the dreaded wartime notices: *Regret advise your son overdue on one of our aircraft yesterday. Further information will be wired as soon as available. Full scale search underway.*

The accident files contain numerous queries from desperate families. One letter to the Department of Transport, dated January 1952, almost six months after the accident, comes from the father of one of the missing men. *Daily I am asked if there is any more news on the disappearance of my son, to which I must reply, "No, nothing more."* He signs his letter, *Ever your grateful and patiently waiting servant.*

A mother states that her lost – and only – son was her sole means of support. She asks pertinent questions concerning the Norseman's airworthiness, its total freight load on that trip, and the qualifications of the pilot who had also paid with his life.

Over the next year, "presumption of death" applications were duly made to the courts so families could pick up their lives, lawyers embark on suits, and the pilot and passengers of the ill-fated Norseman be put to rest, at least legally.

---

* See Chapter 3, "The Unfair Weather of Mount Fairweather", pp 40-48.

A Board of Enquiry heard from numerous witnesses. Ted Williams was apparently *not* in the best spirits that day. He'd told his wife Doris and others that he wasn't happy with the condition of the aircraft and the way it was operating. In fact, he did not want to fly CF-GRQ. "It has no guts," Ted had said, recalling problems he'd had with it on August 2 when leaving Vancouver. The aircraft had shuddered on takeoff and would not climb properly, so he'd returned to the dock. When he went out again accompanied by a mechanic it repeated the problem. The company had given him another aircraft to complete that trip to Zeballos.

Later that day some work had been done on GRQ, including a propeller change after a pilot who'd taxied out the aircraft said the engine was running rough. An inspection revealed that a ring had "come adrift" from the hub, so a replacement propeller was installed. Upon run-up on August 3 the rpm were down, so spark plugs were replaced and further prop work done.

When Captain Len Fraser, QCA's chief pilot, had flown GRQ on August 3, he'd mentioned to both Williams and the mechanic at Zeballos that it seemed to gobble fuel and asked them to check it. But when Williams took it for a flight that afternoon he'd found it to be normal, and expressed no negative comment when Fraser assigned it to him on August 4.

But the aircraft was not operating normally, according to the man who'd serviced it the previous day. "Whether or not they had filled the oil tank in Vancouver that morning, it took quite a bit of oil at noon on August 3 – *eight gallons,"* he reported to the Board of Enquiry.*

The RCAF expressed doubt that weather was a contributing factor to the aircraft's disappearance, especially with a pilot familiar with West Coast flying conditions. Mechanical failure could have brought it down in dense timber, or on one of the few beaches, or into the sea. But if the Norseman had crashed into the ocean or an inlet, wouldn't the prevailing on-shore drift eventually wash parts of the wreckage to shore? None was found.

The RCAF report could only conclude "the crash must have occurred in the trees somewhere between Muchalat Inlet and Tofino." A Board of Inquiry met on November 23, 1951 to state that the results of the investigation were inconclusive.

Bill Cove, who has flown Norsemen and many other aircraft over every inch of this coastline, notes, "The Norseman was an airplane that could easily get away on you during a moment's distraction. If Williams had struck the water on a nose-down attitude in a turn, and cart-wheeled and sank, it is possible that the debris could be swept away in the current.

* The Norseman manual indicates the full oil tank capacity as being 20 U.S. gallons [16.6 Imperial gallons, or 75.7 litres], with the minimum allowable oil reserve as 10 U.S. gallons [8.3 Imperial gallons, or 37.9 litres]. Oil consumption of eight gallons would have been very high, and unusual.

"The outgoing tide could take it past Cape Flattery on the U.S.A. coast. Bundles of logs that have broken from a log-boom have similarly been located down south as far as the Oregon coast," Cove states.

Captain Dan McIvor was the QCA base pilot at Zeballos at the time of the disappearance of CF-GRQ.*

It was to him that Doris Williams handed the logbooks left behind by her husband, as well as two flying reports he'd filled out for trips made on August 2 and 3. McIvor recalls Ted's widow being at a total loss, with her husband missing and three daughters to support.

"The possibility of having an accident while flying that country at that time was always in one's mind," McIvor says. "Any number of things could happen – mechanical failures, hitting a fog bank ..." He thinks the Norseman went into the Alberni Canal and is in there yet.

When given the information that the aircraft had required eight gallons of oil after coming in from Vancouver, McIvor explodes. "If that had happened to me, that airplane would not have gone out! Eight gallons is ridiculous!"

Joe Mugridge, one of passenger Ken Mugridge's sons, still wonders where on the island or in the sea his father's bones might lie. He remains unhappy with the investigative procedures, especially after he became an author of RCAF investigative reports on aviation accidents. He writes:

> *I would have received a stretched neck if I had written some of the 'could-care-less' garbage seen there. That sloppy put-together would never have been tolerated. The gallons of oil used from Vancouver to Zeballos would have caused great concern in the good ole' RCAF. There is also something wrong with the fuel quantities for the Norseman prior to the departure.*
>
> *And two props causing vibration, one after the other, would cause me to check very thoroughly. No one seemed to pay much attention to the pilot's saying that the aircraft shuddered and was sluggish.*

No sign of the pilot, the six passengers, the aircraft or its cargo has been seen since that fateful day. It remains a mystery how the aircraft could have disappeared so completely on this short 37-mile trip. But others who know the coast say perhaps it's *not* so unusual, when flying over mountainous, heavily timbered terrain and a rugged indented coastline, where wind-slashed waves of the Pacific Ocean often wash away man's mistakes.

Jerry Vernon, a B.C. historian and president of the Vancouver chapter of the Canadian Aviation Historical Society, is also aware of the story – as well as rumours that there might have been gold on board. That speculation

* Dan McIvor's biography appears in "The Paper Bag Prince" in *Flying the Frontiers Vol. III – Aviation Adventures Around the World* by Shirlee Smith Matheson. McIvor was inducted into Canada's Aviation Hall of Fame in 2002, and died on February 24, 2005, at the age of 94.

aside, Vernon says, "I have only done a bit of flying in that general area, but the trees are very tall and thick, generally right down to the water's edge. An airplane could just as easily be in the trees as in the water. From time to time, something missing on the west coast of Vancouver Island since the war is discovered when new areas are being logged."

Vernon was a passenger in a helicopter observing some survey work in that same general area, north of Tofino towards Zeballos and Gold River, in the early 1980s, and they decided to take a detour to see where a U.S. Navy Neptune patrol bomber with a crew of 10 had crashed in November 1948 … and was finally discovered "by chance" in October 1961.

"A Neptune is a big airplane," Vernon says, "and the location would have been hard for a large aircraft to get into, unless it came just about straight down. The site is covered with snow and ice most of the year, until September or October, when it starts to snow all over again. It wasn't until the mid- to late 1980s that they were able to get a crew in by helicopter to pick up some more of the remains for a proper military burial. This location is just inland from where the missing Norseman would have passed by.

"There have been so many plane crashes in B.C., and so many aircraft that have gone missing and are yet to be found," Vernon comments. "Several years ago, the Vancouver Island Military Museum at Nanaimo wanted to put up a map of the island with pins stuck in it to represent military aircraft crashes on or around Vancouver Island. I think they expected the number of pins to be modest, but I was able to document 80 fatal crashes, mostly wartime, on or around Vancouver Island alone!"

"Every aviation person on the West Coast is very familiar with these stories," says George Maude, as he indicates the relevant series of clippings in his disaster-filled book. "We all know each other here."

Searching Amidst the Rock Piles

Thousands upon thousands of aircraft complete their flights in perfectly normal fashion, and most passengers relax and enjoy the trip, not considering what might occur if an aircraft should take an unplanned dive into the sea, the rock piles, or forested wilderness.

But mountain search and rescue operators tell interesting stories of hearing an Emergency Location Transmitter (ELT) eerily sound off a mountain, while rescuers circle and circle but find nothing. When the wreckage is finally discovered, often by accident, searchers realize that the beacon sound they'd heard had been deflecting from mountain to mountain, tricking them by camouflaging its source.

"Combine coastal conditions with mountains and you've got a really interesting situation," says Calgary pilot Ralph Langemann. He has flown over most of B.C.'s diverse areas, and came to know its rugged coast while flying in 403 Air Reserve Squadron's search and rescue team. He later flew for Skyway Air Services (Conair Aviation Limited) of Abbotsford, B.C., doing aerial forestry spraying in early summer seasons and fire-bombing in mid-summer.

"The terrain is really rugged," Langemann says, in an understatement. "In some places the mountainside goes straight into the water, there's no beach. You've got a lot of really bad weather, with all that wet stuff coming in. And they're finding out that the old single beacons we used for navigation can be seriously in error if it's wet and there are mountains and there is a coast. They will literally be bouncing off a big wet rock face.

"The pilot might be flying along in cloud and he's got his radio compass pointing exactly where it should be, he's right on course, the whole bit, and he goes smack! right into the rock face, because the signal is not coming to him from the transmitter, it's bouncing off the wet rock face to him."

As a result of current knowledge on how these radio beacons worked, Langemann feels that many coastal accidents written off as "pilot error", or an airplane "just disappearing", could have been caused by the pilot's use of an ADF [Automatic Direction Finder] and honing in on that beacon.

"We'd have to fly into a narrow valley on just an NDB [Non-Directional Beacon] which is transmitting a radio signal. You've got to be pretty accurate to do that," Langemann acknowledges. "You *can*, except that you might not be tracking into the valley at all because the signal might be bouncing off a cliff."

Ex-RCAF pilot Ron Elliot, who later became a flying instructor for the aviation program at Calgary's Mount Royal College (since renamed Mount Royal University), agrees. "A good example of that is when we did all our mountain checks at Invermere, B.C. When we'd be sitting on the ground in the aircraft I'd say to each student that I flew out there with, 'Take a look at the Banff radio beacon.' We knew Banff was one way but the beacon would be pointing another direction. Banff is in one valley, the airplane is in another valley, and the Banff radio station is bouncing off the mountain and coming back at us from 180 degrees."

Then there is the problem of icing, encountered in cloud or rain or wet snow when the temperature gets down near the freezing point, or when flying at altitudes within 5,000 feet above mountain ridges. Water crystallizes to form ice on the props, leading edges of wings, struts, tail, and other exposed surfaces. A pilot must then try to get rid of it either by climbing or by descending if possible to warmer air. Carburetor icing is also a major concern, but the pilot can reduce the engine carb-ice by pulling a lever to add heat.

While there are fewer "mysterious" disappearances or unfound sites in recent years with the advent of radar coverage and electronic search devices, the wilderness mountains, forests, lakes, rivers and ocean still offer many hiding places.

George Maude, Bill Cove, Dan McIvor, and many other pilots can unfortunately recall all too many friends and flying colleagues who've disappeared and not been found to this day. Another of Maude's friends, a 40-year-old pilot, is familiar with the Death and Disasters book, having looked through it a number of times. He's thoughtful as he contemplates the pages filled with clippings, notices, and stories of men who were once like him: young, ambitious, perhaps a little wild, and all infused with a thirst for aviation adventures.

"If I'd have flown when I was younger, I'd probably be in your book," he says.

So many files remain open in Transport Canada's archives – and between the heavy wooden covers of George Maude's book of broken dreams – awaiting factual resolution, emotional closure, and the final clipping that concludes someone's poignant story.

Nazi Toads and Other Wartime Invasions

The Torngat Mountains on the northern tip of Labrador are among the most rugged rock piles in the world. Named for an Inukitut word meaning "home of spirits", these mountains, high-peaked on land, plunge fathoms-deep into the Labrador Sea. Icebergs float down the Hudson Strait toward the Grand Banks in waters that stay close to zero degrees Celsius.

Whales and codfish thrive here, as do char in the fjords, which each summer bring a migration of native fishermen. At lonely Cape Chidley, near the northernmost tip of Labrador, annual precipitation averages 46 cm, half of which is snow. Only the hardiest Arctic flora and fauna survive.

On October 22, 1943, an enemy force in a German U-boat slid through the cold dark waters of this isolated and inhospitable area. Its purpose was to install a weather-reporting station – a daring act that would later bring much interest to Labrador's coast as the only spot on North American soil where the Nazis ever made an armed landing.

How did they manage to do it? And why didn't Canadian, British, or United States military forces, air or sea, record the existence of this weather station? Indeed, this secret Nazi wartime installation remained undiscovered until the early 1980s, when Austrian-born engineer Franz Selinger became intrigued by some mysterious old photographs.

Then living in Ulm, West Germany, Selinger decided to carry out some research on German weather surveys in the Arctic. In the process, he discovered fascinating but vague references to an automatic land weather station in Labrador. This was odd, for all stations but this one had been located at sites in the Barents Sea above Norway. He contacted Canadian military archives, but they had no records of any such establishment on Canadian soil.

Further research put him in contact with the son of Dr. Kurt Sommermeyer, a German scientist who had participated in the automatic weather station program during the Second World War. The son possessed numerous documents, including a series of photographs of an odd type of submarine against an unfamiliar background landscape.

This intriguing material spurred Selinger's intensive investigation into U-boat log books at the German military archives at Freiburg – and his fascination with the anomaly of the mystery U-boat deepened.

Battle of the Atlantic

One week after the *SS Athenia* was sunk off the coast of Northern Ireland on September 3, 1939, Canada, along with the United Kingdom, France, Australia and New Zealand, had declared war on Germany. From that date until the last of the German U-boats surrendered after VE Day in May 1945, the Allied navies and air forces saw no rest. In fact, the Battle of the Atlantic became the longest single campaign of the Second World War. Navigating across the sea lanes in waters infested with submarines and pack ice, or flying through the dense blackness of Arctic nights, with the added challenges of fog, bitter cold and sudden storms, were extremely hazardous. Accurate weather reporting became a necessity to survival – for both the Allies and the Axis forces.

Prevailing weather systems most often travel from west to east across the Atlantic, and prior information obviously favoured the North American Allied forces. Germany therefore was desperate to obtain knowledge of North American weather statistics and provide radio data for their aircraft and U-boats in the Arctic regions.

Since 1920, radio technology had brought isolated Arctic whaling stations into the range of international weather reporting services. German weather stations were established at numerous remote locations, and operated freely until 1940 when the Allies gained control of stations in Greenland and Iceland during further courses of the war.

The Nazis' prior knowledge of the east coast of Greenland, and their ability to establish bases and weather stations in the fjords of isolated ports, were well known. But Labrador and Newfoundland, with bleak, wild, isolated coastlines that were largely uninhabited, were now sought out by German meteorological services (the *Marine-Wetterdienst*, or Naval Meteorological Service; and the *Wetterdienst der Luftwaffe*, or Weather Service of the Air Force) as ideal locations for such covert installations.

Floating weather buoys could be dropped in strategic locations by Luftwaffe aircraft, and unmanned land-based stations could be transported by aircraft or by U-boats, and set up by crewmembers on small rocky islands or peninsulas.

Until the outbreak of war, Newfoundland, then a British colony, had largely been left to its own security devices. Gander Airport and the Botwood seaplane base were set up to sound air raid and coastal security alarms to protect Newfoundland's 6,000 miles of coastline. With its airports serving as important fuelling stops for Pan American Airlines, Newfoundland at that time was more concerned with an American takeover than any threat from German attack.

When Paris fell to Germany on June 14, 1940, the Royal Canadian Air Force took over the Botwood seaplane base "for the duration". Various Canadian commands involving 6,000 men were established throughout Newfoundland, and airfields were set up at Torbay and Goose Bay, Labrador.

On September 2, 1940, the Americans moved in under a Lend-Lease agreement with Britain to supply destroyers and other war items in exchange for 99-year leases on bases in Newfoundland and other places. Thus, American and Canadian forces performed joint commands at Torbay and Goose Bay, eventually becoming 10,000 strong throughout Newfoundland and Labrador.

The Battle of the Atlantic grew in intensity as the Allies struggled to sail their convoys safely past U-boats that plied the waters, and thwart the enemy's attempts to set up bases on the east coast of Greenland. In order to alleviate the effectiveness of such Nazi bases, on December 3, 1941, Washington approved plans to establish a new air patrol base in the North Atlantic.

The United States Army Air Force chose the Norwegian island of Jan Mayen between the Greenland and Norwegian seas, northeast of Iceland and half way between there and Spitsbergen, as the location for its base. The little island, once a whaling centre but now a mountainous wasteland, supported an antiquated Norwegian weather station that had been set up in the 1920s. The Americans believed this site could also be used as a refuelling base for ferrying aircraft to Archangel, Russia, so its rudimentary airport was made available to the U.S.A. through agreement with the anti-Nazi Norwegian government-in-exile. Also, air patrols could fly out of Jan Mayen, where they would be ideally positioned to intercept Nazi supply ships coming from conquered Norway, occupied by Germany from 1940 to 1945.

A Proliferation of Toads

Meanwhile, German engineers were busy developing futuristic weather-reporting systems for manned and unmanned weather stations, considered essential to safe navigation of their aircraft and U-boats in the North Atlantic.

Patented land-based automatic weather reporting stations called *Wetter-Funkgerät* [Weather Transmitter] *-Land [WFL]* and *-See [WFS]* were developed and manufactured in 1942 by Siemens-Schuckertwerke AG Berlin. The project was led by Defence Research Group Professor Dr. E. Ploetze and Head Engineer E. Weidemann in co-operation with the Nautical Science Group of the Naval High Command.

While the German Navy was transporting and installing these weather stations in remote sites, engineer F. Woelfle of the *Reichsamt für Wetterdienst* [Meteorological Service of the Reich] was designing automatic weather stations called *Kröten*, or Toads, for the Luftwaffe. A storm-resistant sheet metal weather box held sensors for air pressure, temperature and humidity, along with a probe, transmitter and timer. An antenna was stretched five to six metres high between two masts.

Franz Selinger describes this venture and the challenge to keep the Toads operational:

> *The first Toad was flown to Spitsbergen in May 1942 in a Heinkel He 111 piloted by Rudolf Schütze and set up on a beach in Advent Fjord, where over the summer it was to bridge the gap between the evacuation of the manned Luftwaffe weather station which had been working in the vicinity since November 1941 and the station's reoccupation in the fall of 1942. By July of 1942, however, this station had already been discovered by a commando unit of the Norwegian garrison in Spitsbergen, dismantled and shipped to England.**

The next Toad was delivered by air to Bear Island (the southernmost island of Svalbard, in the Norwegian archipelago) in October 1942. A work detail had to be parachuted in to prepare a landing field for the Heinkel He 111. The twin-engined monoplane, typically flown as a bomber, carried a crew of five, 2,500 kg of bombs, and was armed with five machine guns and one 20mm cannon. It was initially used in Spain in 1937, and then throughout the war. Later, it was consigned to night bombing, mine-laying – and Toad-placement missions. The following spring, a German navy landing force found this Luftwaffe station on Bear Island still in good condition, although no longer operating since storms, or polar bears, had broken the antenna masts.

---

* No record apparently remains of what became of the station after it reached England.

The third Toad was planned for Mezhdusharskiy Island offshore Novaya Zemlya, part of an archipelago north of Russia between the Barents and Kara seas that formed part of the Soviet Union. The plan failed when the Heinkel He 111 was detected on the island by a Russian seaplane from Belyush'ya Guba, and the operation was cancelled.

In co-ordination with the navy, the Luftwaffe Meteorological Service scheduled additional operations in 1943 on Spitsbergen, Edge and Bear islands. They also planned to develop *Land-Kröten*, stationary automatic weather stations that could be dropped by air. These would be set up on islands between Jan Mayen and Novaya Zemlya and throughout Arctic regions, with *See-Kröten* floating on the Barents Sea.

An Arado Ar 232-A tactical transport aircraft, an all-metal high-wing monoplane used by the German military, was specially modified for these flights. In addition to retractable landing wheels under the wing, and a nose landing wheel that retracted backwards into the underside of the fuselage, there were 11 pairs of small auxiliary fixed wheels beneath the fuselage that allowed the aircraft to make emergency landings on rocky terrain (with its row of fixed wheels the Arado bore a certain resemblance to a millipede, which inspired the German name for the aircraft: *Tausendfussler*).

However, even the Arado Ar 232-A could withstand only so much beating, and the aircraft sustained undercarriage damage in July 1943 while landing on the south cape of Spitsbergen. After setting up the Toad, pilot Rudolf Schűtze and his crew were forced to make emergency "wilderness" repairs, and take the aircraft back to Banak Air Station in Norway. (During the Second World War the German occupying forces expanded this airport in Norway, with hangars, workshops and hospitals in addition to defensive positions. The buildings and installations were destroyed during the German retreat in 1944. This station remained virtually forgotten until Franz Selinger spotted it during a reconnaissance flight in 1984. While some damage had been done to the station "at the hands of persons unknown," the main components remained there until salvaged in 1985.)

In August 1943, a few weeks after the south cape landing, the same pilot and aircraft transported another Toad into the Diana Valley on Edge Island [Edgeøya], an uninhabited Norwegian island lying southeast of the Svalbard archipelago, east of Spitsbergen Island. (This station was re-discovered in 1969 when a Norwegian Polar Institute helicopter spotted the cleared runway tracks on the valley floor and landed to take a closer look. It was salvaged by the Norwegian Coast Guard in 1984, to be restored for the Defence Museum in Oslo.)

Schűtze, a pilot regarded as a pioneer of takeoffs and landings in impossibly rugged northern zones, delivered a further station to Bear Island in August, 1943. Bad weather grounded the crew there for two weeks. They were finally forced to jettison fuel and lumber intended for constructing a log road in order to take off from the waterlogged landing field. On the return flight homeward after leaving northern Norway, the venerable Arado Ar 232-A experienced engine failure in bad weather and plunged into the sea. The entire crew was killed. The three stations they had set up continued working until later that fall.

This setback was a severe loss to the Nazi plan to establish automatic weather stations in the Arctic. Use of aircraft for this important strategy had freed the expensive submarines for combat purposes, but Schütze's expertise was integral to the success of the operation. Only one more Toad was deployed by the Luftwaffe, at the end of 1944 on the Alta Fjord in northern Norway, when the area was evacuated.

A Pack of Wolves

Following his retirement, engineer Franz Selinger began compiling information for a book about the German weather stations in the Arctic. He calculated that Germany had actually built seven *Kröte* stations for operation by the Luftwaffe, and 16 *Wetter-Funkgerät-Land*, or WFL, for set-up by the *Kriegsmarine*. As well, a number of other types of weather stations were operated by the navy or the Luftwaffe in Arctic or sub-Arctic regions (Spitsbergen, Bear Island, Franz-Joseph-Land and Greenland), and five around the Barents Sea above Norway. Two more were planned for North America: WFL-26 (called "Kurt" after Dr. Kurt Sommermeyer) and WFL-30 (named "Herbert" after U-boat commander *Oberleutnant zur See* Herbert A. Werner) to replace WFL-26 after it had stopped transmitting for good.

By the spring of 1943 the Germans had as many as 235 U-boats roving in wolf-pack teams of five or more, intent on attacking convoys in the North Atlantic Ocean. Knowledge of weather conditions in the northwest Atlantic was imperative for their success.

Newfoundland, even though protected by anti-submarine warfare measures, was acutely aware of these wolf-pack attacks, especially in the "black pit", a gap in the Allied defences midway between its eastern coastland and Iceland. A number of attacks by U-boats hit Newfoundland-area targets, killing crews and passengers on ships ranging from Canadian ore freighters to the Newfoundland Railway ferry *SS Caribou.*

With the increased vigilance by Allied forces to track these attacks and launch counter-attacks, it seems impossible that U-537 could cross the Davis Strait, take a bearing from Cape Chidley on Killinek Island near the northern tip of Labrador, manoeuvre down the coast to drop anchor off the rocky shoreline at Attinaukjuke Bay [now Martin Bay] on the Hutton Peninsula, at 60°5' N, 64°24' W, and then, on October 22, 1943, set up a land-based, battery-powered, weather transmitting station. But it did.

The Mysterious "Kurt"

Dr. W.A.B. "Alec" Douglas served for 23 years in the Royal Canadian Navy, and later became an historian for the Canadian Armed Forces after co-authoring *Out of the Shadows: Canada in the Second World War*. He was quite unaware of any weather station in Labrador until he received correspondence in 1980 from Franz Selinger.

The records were all there, Selinger said, housed in German military archives in Freiburg, and included the meticulously kept logbook for U-537. This U-boat, a Type IXC/40 built in 1942, was initially constructed for regular operations. But then, under the direction of *Befehlshaberder Unterseeboote* [Commander U-boats] Admiral Karl Doenitz, the logbook indicated that U-537 was set to sail for Canada, with a mission to transport and install land-based automatic weather transmission station *Wetter-Funkgerät-Land Number 26*, or WFL-26.

U-537 had undergone the usual naval training with the 4th U-boat flotilla at Stettin, Germany. But when its destiny was altered with the assignment of the "special mission", the U-boat was to be equipped with a 20mm quad anti-aircraft gun mounted on the after-section of the bridge as a reinforcement of her anti-aircraft armament. On July 25, 1943, however, shortly before this modification was completed, the U-boat was damaged in an air raid, necessitating a holdback in the dockyard while storage batteries were replaced. By September 11 the batteries were in place and U-537 was ready for operational patrol.

Crew members were not aware that their upcoming patrols would be extraordinary until a government administrative advisor [*Regierungsrat* Dr. Kurt Sommermeyer] and a nautical assistant [*Nautischer Assistent* Walter Hildebrandt] came aboard.

Werner Bendler, a crew member on U-537, noted in his memoirs, "The crew was surprised that specially packed instruments were stowed away in the upper deck pressure compartments which normally housed the torpedoes, and these were under the supervision of two government officers. ... It came as a great surprise that these two made the journey with us, as this was most unusual."

On September 18, 1943 U-537 sailed from Kiel, Germany under the command of *Kapitänleutnant* [Lieutenant Commander] Peter Schrewe, to cross the western North Atlantic. Following a propeller shaft repair, and a stop to take on provisions and fuel at Bergen, Norway, U-537 departed on September 30 for its special mission.

As Bendler describes the voyage:

The journey now went in the direction of the Shetland and Faeroe islands, and not until we reached the open sea did the commander dare to open a sealed envelope in which his instructions were written. The duties consisted of the building of a fully automatic weather station in Labrador, Canada, consequently on Canadian soil – and all this was in the middle of the war.

Any contact with the enemy was now forbidden in order to safeguard this powerful undertaking, and in order to keep the highly sensitive instruments out of danger.

Heavy seas near Cape Farewell on the tip of Greenland ripped the U-boat's anti-aircraft gun from the outside of the boat, and the attack periscope failed because of water leakage. Orders were to simulate the presence of a U-boat wolf pack by transmitting signals from various positions, but to avoid detection by remaining underwater as much as possible. By October 18 the U-boat was again on its way to its destination: the coast of Labrador.

A translation of *Kptlt.* Schrewe's war diary states:

Position AJ 2874, wind NW7, sea 6, overcast, visibility 12 nautical miles, 1009 mb. To avoid if possible any contact with humans – for instance, with Eskimos, who at this time of year travel south – I will choose for the erection of the radio weather station a place lying on the northern part of the coast of Labrador. Target square is AA 1499. Ice is not to be expected yet. Backing off in a southerly direction remains an option.

Seeking the most northerly and remote spot they could reach on North American soil, *Kptlt.* Schrewe directed the U-boat across Davis Strait between southeast Greenland and Baffin Island to Cape Chidley and south along Labrador's coast. It was necessary to take continuous depth soundings in this largely uncharted region to avoid the many reefs and ledges.

Bendler states in his memoirs that they were guided by aerial photos and precise maps. However, Franz Selinger questions the idea that the U-boats might have made use of aerial photographs or exact maps:

If ever aerial photographs of that region existed at that time, they could not reach Germany to serve U-boats for landing actions. The sea charts used by German U-boats were mostly based on pre-war issues of the British Admiralty Charts, and even German survey vessels that surveyed foreign coastal waters in peace time never steamed up Davis Strait to 60° North. I do not know what chart Schrewe had to draw the route sketch later enclosed in the log book, but he found his way to Martin Bay by visual observation and careful sounding in the channel between Avayalik and Oo-oililik [Home] islands and [rounded the southern tip of] Hutton peninsula.

Bendler recalled that the U-boat came in between Home and Avayalik islands and anchored "with the 10.5 [cm] gun and the torpedo tubes pointed to the inlet entrance in order to be ready for any surprises." The ship ran south to the mouth of Martin Bay and anchored on October 22, 1943 at 18:45 hours, just inside the southeast tip of land some 300 metres from shore (60°4.5' N by 64°23.6' W).

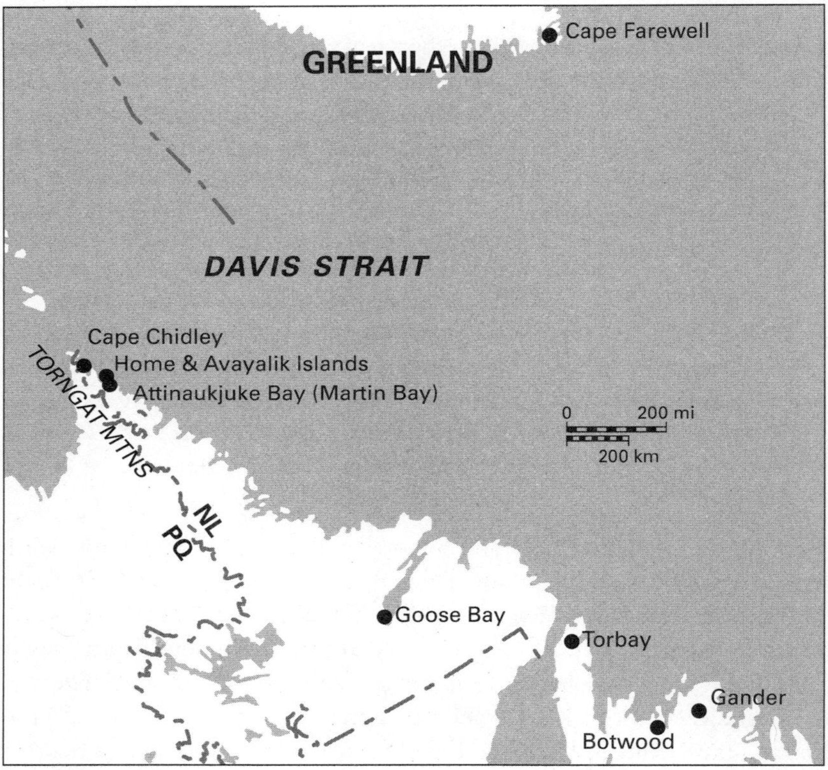

The Intrusion of U-537

A patrol was sent to explore the land area while Dr. Kurt Sommermeyer advised the crew about unloading the measuring instruments from the pressure containers, along with the battery equipment, rubber dinghies, outboard motors and a variety of armament devices. Ten crew members loaded the components in the dinghies and waited for the fog to roll in while a core crew stayed alert on the bridge, armed and ready for action if spotted. They were completely at risk as the U-boat could not submerge in the shallow water, and escape was quite impossible.

A lookout posted on the seaward side of the ridge on the peninsula sighted a patrol boat and gave warning. The crew of U-537 immediately went on high armed alert, but the patrol boat went by without entering the inlet.

The U-boat crew's task was Herculean, and dangerous. This remote territory was in enemy hands, far from Axis assistance. Even if U-537 remained undetected, radio transmission from the region could quickly result in discovery, followed by an attack.

Work continued under instructions from the electronic and meteorological engineers. A nine-metre tower was built to transmit the atmospheric surveys. Around the tower, and anchored to it, was a circle of 10 grey-painted canisters, each weighing 100 kg and containing nickel-cadmium dry-cell batteries; when one battery was used up or failed to operate, the next one would kick in. Accompanying these were the transmitter and weather-measuring devices; it took four hours to assemble and test the station.*

The designation "WFL-26" was painted on the transmitter canister in dark grey or black lettering. Further, they crudely inscribed the words "Canadian Meteor Service" in white letters, believing that an organization with such a general name would surely exist. These markings were intended to simulate the Canadian origin of the station, as (they hoped) a plausible ruse to prevent their destruction or removal by Allied visitors. To avoid suspicion if the site was discovered, the crew then scattered empty American cigarette packages and match folders about the ground.

At 18: 00 hours on October 23, after completing a job that had taken just 28 hours, and ensuring that the station was functioning properly on its frequency of 3,940 kilohertz, U-537 quietly weighed anchor and headed out to sea to reach a position approximately 500 km from the coast. There, on October 25, they radioed a message regarding the completion of the "special assignment".

When in operation, WFL-26 would be automatically activated by a timing device. Every three hours a radio transmitter would make a three-minute report (one minute for warm-up, two minutes for transmission) that included temperature, barometric pressure, wind force and direction.

The battery-powered transmitter broadcast on the high frequency band, between 3,000 and 12,000 kilohertz. Although it had an output of just 150 watts, this would give it sufficient range to reach Europe. The first transmission of WFL-26 occurred three minutes late, but otherwise was technically perfect.

---

* Franz Selinger describes the apparatus that held the probes, the transmitter/timer unit and the batteries as 10 barrel-shaped steel containers, 520 mm in diameter and about 1 m high. The 10-m-high rod antenna stood on a tripod. Some of the battery containers were placed on the feet of the tripod to provide increased stability. A short mast held aloft a wind vane and a cup-type anemometer, set to operate on fixed frequencies between 3.8 and 10 MHz. Meteorological values converted to Morse code were sent four times a day at predetermined times to receiving stations, with the information being sent on to a central weather office for evaluation.

Air attacks

U-537 was then free to carry out an anti-shipping patrol off Newfoundland's coast. Bendler describes their simulation of various positions of non-existing U-boats by sending radio communications in order to attract anti-submarine aircraft and naval patrols and thereby tie up enemy strength so other German U-boats might fulfill their missions undisturbed.

"This was suicidal for us," states Bendler, "since we were, of course, being hunted in a stationary position. This almost led to our destruction."

Obeying orders from U-boat headquarters to observe convoy traffic and patrol within 250 km of St. John's, Newfoundland, U-537's position was revealed through Enigma decrypts by Allied forces on October 29, 1943. Canadian air and naval forces immediately mounted an exhaustive hunt (code-named "Salmon") to find and destroy it.

Although U-537 was equipped with a Naxos radar detector, it could not avoid detection and came under fire on October 31, the third day of the "Salmon" search. A Lockheed Hudson from 11 Squadron, RCAF, piloted by Flying Officer F.L. Burston (whose brother was the navigator), found the U-boat floating on the surface and fired eight rocket projectiles, the only time such armaments were used in action by an Eastern Air Command aircraft. No direct hits resulted.

Dr. Douglas states:

> *The rockets, with solid steel semi-armour-piercing warheads, had been delivered to 11 Squadron and 119 Squadron the month before. The Hudsons, fitted with eight of these projectiles, were supposed to aim 20 yards short of the submarine to hole it below the waterline, and in this instance the pilot probably undershot. No. 1 Group dispatched or diverted three aircraft to carry out a Salmon, but the Hudson, its crew unaware of the importance of continual air coverage to keep the U-boat down and immobilize it until naval forces could reach the position, left the area after making its unsuccessful attack … Mystified as to how he had been located and taken completely by surprise, the U-537's captain concluded that "this part of the coast has been made unhealthy," and retreated to the south.*

Still patrolling the Canadian waters, however, and checking on his weather station, *Kptlt.* Schrewe noted in the U-537 logbook on November 4 that there was intensive jamming on the frequency.

At 21:30 hours on the evening of November 10, approximately 325 km south of Cape Race, U-537 again came under repeated aerial attacks from a Canso flying boat of 5 Squadron, RCAF, that was escorting convoy Halifax 265. The Canso, attacking through flak, dropped four depth charges, all of which fell wide.

"Admiral Murray promptly ordered another 'Salmon'," Douglas writes, "but once again everything went wrong. The Canso thoroughly confused the warships escorting Halifax 265 by incorrectly reporting the position of the attack, and then departed before relief aircraft arrived. As a result, U-537 was able to escape by making a fast run on the surface for five and one-half hours."

The next morning, however, Pilot Officer R. Duncan, flying another Canso from 5 Squadron, spotted U-537 at a position 80 nautical miles farther south. He managed to drop four depth charges, even while his aircraft suffered a large hole in the leading edge of one wing from flak from the U-boat. The Canso's depth charges exploded just 20 metres off the U-boat's bow, causing it slight instrument damage.

A Canso that came in to relieve Duncan's called in warships to the area, led by the British destroyer *HMS Montgomery* (formerly USS *Wickes* but re-commissioned by the Royal Navy as part of the U.S./Britain "Destroyers for Bases Agreement"). However, by this time U-537 had submerged. The ships searched a 15-mile radius from the position of the attack, assisted by two Cansos from 5 Squadron, but intense fog rendered the search futile. While the ships continued the hunt for three more days, later accompanied by more Cansos from 5 Squadron and three Liberators from 10 Squadron, the hunt proved unsuccessful.

Again checking on his weather station installation, *Kptlt.* Schrewe wrote on November 18, "Frequency of weather station is jammed by unknown own [German] radio station. Weather transmissions not received at scheduled times."

On November 19 a signal from U-537 was picked up by Canadian shore stations. No. 10 Squadron again sent out Liberators to make four sweeps of a suspected area toward the mid-Atlantic – a full week before U-537 finally slipped away from the North American coast under orders for "free pursuit". The "Salmon" hunt was considered to be a dismal failure.

With its fuel low, the warrior submarine sailed for home.

While in the middle of the Atlantic, however, news reached the U-537 captain that the enemy was in the process of locating their position with a new type of range finder; the U-boat would have to keep its radios silent. One of the engineers aboard ingeniously constructed a temporary arrangement that allowed the U-boat to determine enemy bearings so it could promptly submerge and escape. "And so it was that this man with his primitive inventiveness brought us out of distress and doubt, and saved all our lives," writes Bendler.

U-537 docked at Lorient in the Bay of Biscay in France, the base for the 2nd and 10th U-flotilla, on December 8, mission accomplished. While Bendler states that Weather Station "Kurt" worked for 18 months, other sources believe it worked for a much shorter period of time before the batteries were exhausted.

The next assignment for U-537 was to sail to Singapore for a mission in the Pacific. Hildebrandt and Dr. Sommermeyer, as well as crewmember Bendler (who was called to attend non-commissioned officers' school), disembarked before the submarine's next voyage. On November 9, 1944, U-537 was torpedoed and sunk in the Java Sea east of Surabaya by U.S. submarine *USS Flounder*. All 58 crewmembers drowned.

The information provided by the Labrador station must have proved useful, however, because in 1944 Admiral Karl Doenitz sent out U-867 to set up a second automatic weather station (WFL-30, or "Herbert") on Labrador's coast. But this submarine was sunk en route by RAF aircraft, and no further attempts were recorded.

Why Was "Kurt" Never Discovered?

Weather station WFL-26 remained one of Germany's best-kept small secrets, known only to a few scientists and seamen until research by engineer Franz Selinger brought its existence to light. He had become intrigued when, among Dr. Sommermeyer's papers, he found photographs of the unusual U-boat lacking a quadruple anti-aircraft gun and sitting at anchor against an unfamiliar configuration of hills.

The weather station depicted in these photos also was quite different from eastern Arctic installations located on Greenland and Svalbard, the Norwegian-owned group of islands forming the Spitsbergen archipelago. The location wasn't Spitsbergen or Bear Island, or other locations where unmanned weather stations had been established. Selinger consulted German naval historian Jürgen Rohwer to learn that the mysterious U-boat in the photographs was U-537. Selinger scrutinized the U-537's logbook and learned that the submarine had visited Labrador. Could that be the site of the unfamiliar coastline?

In 1980, Franz Selinger wrote to Dr. Douglas, inquiring what information the Canadian military archives might have on a weather station in Labrador.

There was none.

Neither could the American authorities provide any information. It was impossible, they said; radio transmissions would have brought air patrols on the double. Nothing had been reported.

Sceptical but interested, Dr. Douglas requested records from Operations Intelligence Centre at naval headquarters in Ottawa, but was informed these documents had been destroyed after the war. Further research of files from the Flag Officer Newfoundland, the Commander-in-Chief Canadian Northwest Atlantic and Eastern Air Command of the RCAF, revealed frequent references to U-boat sightings, but none on the Labrador coast that had been intercepted.

Dr. Douglas turned his search to the Canadian Coast Guard and to its director of fleet systems, Captain J.Y. Clarke, a former director general of the Transport Canada Training Institute in Cornwall, Ontario. Intrigued by their discovery, Clarke invited both Douglas and Selinger to make a site visit that summer aboard the icebreaker *Louis S. St. Laurent* on its way from Halifax to Baffin Island.

Mysterious Intruders

On July 21, 1981, they sailed on the icebreaker as close as possible to Martin Bay, and then flew overland in the ship's helicopter to find the site. With copies of the wartime photos in hand, they soon identified landmarks. There was the old weather station!

Elation was crossed with disappointment, however, because the site had been extensively vandalized. Canister after canister had been opened and the contents strewn about. But this didn't resemble "senseless" vandalism. Radio parts and batteries had been uniformly disconnected, and cables cut and removed. Identifying plates on various components had been obliterated. Parts of the coding device were there, however, as well as the tripod and mast, and some dry-cell batteries. Based on the old photos, they realized that all but one of the canisters were still onsite.

Bending down to examine the rocky ground, Selinger discovered a .303 rifle cartridge marked "British Dominion" beneath one canister. Someone had destroyed the station and attempted to obliterate markings of ownership. But who?

When Canada and the U.S. were establishing DEW [Distant Early Warning] stations in the North, the sites of Home Island, as well as Saglek Bay in Labrador, had been visited in 1952 by the U.S. Navy's hydrographic team. No apparent record remains as to why the U.S. Navy chose these two little dots on the map to locate weather stations. Nor do official records exist of their finding the Nazi weather station.

Professor J. Peter Johnson, a geomorphologist at Carleton University, Ottawa, reported on his visit to the site in 1977 while shooting a traverse for a geological survey. He hadn't then realized its significance, but recalled that the now-missing cables had been intact, although the other materials were in the same wrecked state as found later in 1981.

The establishment of the Martin Bay weather station coincides directly with a crucial phase of the U-boat war; thus the find is significant. The technical and physical expertise associated with the exercise is also surprising, as the Canadian Atmospheric Environment Service [AES] did not utilize such structures until the early 1960s. Although a number of automated weather stations now operate in Canada's North, Dr. Douglas notes, "It is doubtful if any of them could have been packaged into 10 cylinders weighing no more than 220 pounds each, and capable of being loaded into and unloaded from a conventional World War II submarine. The need for accurate weather information drove the Germans to extraordinary efforts. We now know the extent of the risks they were prepared to take with precious U-boats to acquire this information from our northern regions."

The last manned German weather station of the Second World War, code-named *Haudegen*, was installed during the summer of 1944 at Svalbard in Northeast Land. There, geographer Dr. Wilhelm Dege and a crew of 11 men set up residence for nearly a year, from September 14, 1944, to September 5, 1945, to collect and transmit weather data and to explore and map the ice-free corridor extending south across the area. The crew was forgotten when Germany surrendered in May 1945, and was finally rescued on September 4 by a Norwegian sealer. They were the last members of the German armed forces to surrender at the end of the war.

The Risks

Franz Selinger sums up the reasons Germany went to such lengths to risk the lives of pilots, sailors, meteorologists, explorers and geographers, as well as expensive airplanes, ships and equipment, to install these weather stations:

> *Whereas allied meteorological services could maintain manned weather stations in all important areas without risk and also had access to weather reports from the numerous ships at sea, thanks to their almost unlimited air and naval supremacy, the German command was forced to look for ways and means of obtaining its own reports from these same areas. This could only be done with a few manned stations set up in hard-to-reach locations, and with the automatic stations. By the end of the war, German development work in this area had reached a level which at that time was unique. It was only years later that stations of this kind would come into use in other countries.*

What seems odd is that the achievements of these men and machines in developing and deploying the automatic weather stations during the Second World War came to the notice of the North American public only after Selinger's research nearly 40 years later into the Labrador-based station.

Curiosity about this station did not end with Selinger's and Douglas's voyage of discovery in 1981, however. Environment Canada's senior climatologist, David Phillips, has collected relevant newspaper clippings and magazine articles and a number of academic papers that focused on the topic.

Phillips also conducted an oral history session with Dr. Patrick McTaggart-Cowan, director of the Meteorological Service of Canada 1950-1964. As McTaggart-Cowan was the officer in charge of the weather station in Botwood, Newfoundland during the war (his title was Chief Meteorological Officer, Western Atlantic Region, RAF Ferry Command), he would have been the person responsible for meteorological briefings during the North Atlantic air lift.

When Phillips asked about the foreign weather station, he was surprised by Dr. McTaggart-Cowan's response, which seemed to dismiss the importance of WFL-26: "We had some incredibly good radio officers in Newfoundland at that time. There were more important priorities. It was a momentous year in the Battle of the Atlantic. It was not worth going out of our way to destroy one small German transmitter. We knew its broadcasts would not last that long."

But a former head of automatic instrumentation for AES begs to disagree, pointing out that the German Labrador station stood out not because of its design but because of its longevity. "Those batteries were remarkable," he reported to *Zephyr*, the staff journal of the Meteorological Service of Canada, Environment Canada, 1990. "It was only after Mars II in the late 1960s that Canada had anything comparable.* Its solar panels allowed long-term transmissions, so automatic stations could be left unmanned in the remote Arctic."

David Phillips notes, "Dr. McTaggart-Cowan added that unfortunately records of the station were destroyed after the war and that he had neglected to tell his weather service colleagues about 'a fairly minor wartime episode.'"

Dr. McTaggart-Cowan also stated that he was the only Canadian with Ferry Command, so it was unlikely anyone else in Canada got to know about it. He commented that the technology was "pretty elementary," and downplayed its accomplishments when compared with those he regarded as more noteworthy. "If we want to remember war operations in the North Atlantic, why not recall the building of Goose Bay airport? It was a feat all Canadians should be proud of."

Despite these contrary opinions, one cannot help but be impressed by Germany's development, transport and installation in enemy territory of the land- and sea-based automatic weather reporting stations, *Wetter-Funkgerät-Land* [*WFL*] and *-See* [*WFS*].

"Whatever the final destiny of the captured weather station [at Martin Bay], it will always be an outstanding illustration of how far a country

---

* Mars II was the collective name for a planned permanent space infrastructure for exploring Mars. Included in the concept was a plan for in-situ resource extraction from the surface of the planet. Although the program ultimately failed due to various system malfunctions, Mars II, together with the parallel venture Mars III, significantly advanced technology, and did in fact succeed in placing the first human artefacts on Mars.

involved in total warfare can go to mount a daring and lonely mission deep into enemy territory for the sole purpose of obtaining accurate weather information," notes the article in *Zephyr*.

This and other stories, now over 60 years old, are still discussed when recalling daring and incredible tales of war. Rumours abound, attaining the status of folk myth. Some say that German U-boat crewmen even made numerous visits to the rocky shores of Newfoundland and Labrador.

"There are still many questions left unanswered," writes Dr. Douglas in his article titled *The Nazi Weather Stations in Labrador*.

He thoughtfully adds, "They may never be."

Soaring with Seagulls

Providing a safe haven to an amazing variety of birds, Grand Manan Island, New Brunswick is on the eastern flight path for more than 400 migratory species. More recently, it has provided a sanctuary for a tenacious aviation enterprise.

Klaus Sonnenberg's fledgling fleet, named "Seagull" by a friend who listened incredulously to Klaus's plan to start an amphibious flying service on the island, emulated many of these birds' attributes in order to survive. It required him to be resourceful, inquisitive, intelligent and determined. For, like most other settlers who came ashore, either through choice or chance, the airline's owner faced significant challenges from the very outset.

Grand Manan was in desperate need of pilot Klaus Sonnenberg's services. The island, populated by some 3,000 persons, measures only 24 km in length and 10 km at its widest point, and guards the entrance to the perilous waters of the Bay of Fundy. Although beautiful with its cliffs and crags, the island is often shrouded in fog that thickens the skies and screens the dangers of the surrounding shoals and tides. Its principal link to the mainland had always been the ferry, but by the early 1990s this system no longer met the needs of residents, especially during emergencies.

Klaus Sonnenberg had realized early in his aviation career that to win flight he often had to fight, and in 1981 he was fit and ready. Klaus and his wife Melanie went door to door with a petition seeking support for their application to Transport Canada for a commercial air service based on Grand Manan Island, and to their delight *all* residents of voting age signed it.

First flights

Klaus Sonnenberg was born in Germany on January 1, 1946, and inherited the aviation bug through stories told by his father, Siegfried, who had been inducted into the Hitler Youth group as a boy. "The Germans had a Youth Flying Club, you might call it, and they built airplanes and flew them," Klaus recalls. "When the war started he was already an advanced machinist/mechanic, so they put him in charge of an attack group."

After the war, the family moved to Canada and operated a resort on Kootenay Lake in British Columbia, where Klaus came to know and love the outdoors. Earning a Bachelor of Science Degree in Zoology from the University of Idaho, followed by graduate work, he took employment with the Alberta government in wilderness education. He then moved across the country to the Maritimes. When he first started flying lessons, in 1975, Klaus was Chief Park Naturalist in Cape Breton.

Eager to master this new skill, he realized if he was going to become proficient in flying he should have his own airplane, and in early 1977 he purchased a Piper Tri-Pacer. When his licence arrived on May 20, 1977 after a long and impatient wait, he set out on what was to be the first of many adventurous flights, to Saint Pierre and Miquelon.

These French-owned islands in the Gulf of St. Lawrence are comprised of a barren, rocky landscape but remain a hotly disputed territory between Canada and France, mainly because of fishing rights, and currently offshore oil rights as well. "I just had a desire to go someplace exotic," Klaus says. "I had no idea about bad weather flying, and I got into freezing rain." Not a great inaugural trip, but Klaus's next flights were even more ambitious and further afield.

On July 30, 1977, he flew over 260 km of open water to Sable Island. With dimensions of 42 km in length and no more than 1.5 km across at its widest point, Sable Island resembles a large snake curving along the ocean's continental shelf. This time, however, Klaus was prepared to encounter some bad weather, having learned something about the island's history. Once referred to as The Graveyard of the Atlantic for its frequently foggy conditions due to clashes between the cold air currents from Labrador and the warm Gulf Stream, and sudden storms that often included hurricanes and nor'easters, the shoals surrounding Sable Island were littered with over 350 shipwrecks, many visible from the air.

The Nova Scotia government in 1801 had established a lifesaving station on the island, and only this crew was allowed residency. The severe restrictions regarding access to Sable Island had worked to prevent plundering of the many shipwrecks, and more recently to protect the island's fragile habitats. Klaus also knew that in 1960 the federal government had placed this island under environmental protection to save its wondrous variety of flora and fauna, including a world-famous herd of over 300 wild horses known as the Sable Island ponies. Written permission to visit had to be obtained from the District Marine Agent of the Department of Transport in Dartmouth, Nova Scotia, and was usually given only for emergencies or to federal government employees or research scientists whose duties required a brief stopover, or for longer work-related habitation of the island.

Thus Klaus's call to gain official permission was met with a blunt response: "You're crazy. You have to write us a letter, and in a month or two we may write you back."

Klaus was undeterred. "That's difficult, because I'm with Parks Canada and we've already chartered this aircraft," he said. "As a government body I'm asking you to expedite this." A small hesitation was followed by a big sigh. "You government department guys are all the same, you never follow the rules." Then, "Tell me what air charter company you've hired. What type of aircraft?"

"I don't know," Klaus replied. "I do have the letters of the aircraft, though: C-GLDW."

Klaus had to be cautious, knowing that if he'd admitted it was his own private Tri-Pacer the conversation would come to a sudden end. "I was told that all you needed was the identifier."

The application for approval went ahead. However, Klaus had no real navigation equipment on board, just an old combination NavCom that really wasn't up to par. He and a passenger left Sydney at about 9,000 feet over the clouds, trying to stay tuned in to the VOR [Very High Frequency Omni Range] signal out of Sydney as long as he could. "Once we lost the VOR – we had no ADF [Automatic Direction Finder] – we went by dead reckoning until we broke below the clouds. Then I said to my friend, 'We're low on fuel and there's a lot of haze. If we don't spot this island in two minutes we'll have to turn around and go back to Sydney.' We had extra fuel in the baggage, but of course to put it in we'd have to land. We couldn't land, we were over water, and we'd run out of fuel if we continued."

They were about to turn around when just ahead of them was Sable Island. They circled the island, looked down at the shipwrecks, seals, and wild horses, and then picked out a wide spot to land on the beach near to where the environmentalists lived. Local residents came to meet them on a tractor and, because the visit was considered official, with full permission granted, they didn't ask too many questions.

"They wondered, I guess, how official this was, but nevertheless they fed us lunch and then took us around in the tractor and we got wonderful pictures of exploring the island," Klaus says. "Then we filled up with gas and away we went." In the meantime the weather had turned sour, but, flying at 200 to 300 feet altitude, they finally made it back to the airport in Sydney.

On August 4, 1977, with no more than 50 hours airtime experience, Klaus flew right across Canada to Long Beach on the coast of British

Columbia, and back. "I was working for Parks Canada, so I arranged with the Park Warden to land at Long Beach for lunch."

Back home, Klaus set out to build his own airstrip in Cape North, near the village of Dingwall in northern Cape Breton. He took a government grader from the Parks Department and levelled out the strip from behind the senior citizens' lodge. "I had to build a corral to protect the plane because there were cows there," Klaus notes. "The plane was fabric-covered and the cows would destroy it. That's where I flew out of for several years."

His first forced landing occurred on May 19, 1978, at Port Hood on the southwest corner of Cape Breton. Klaus explains, "The engines on the old Tri-Pacer came with 7/16-inch exhaust valves, which had an AD [Airworthiness Directive] that stated you couldn't fly them more than 1,200 hours without changing them to ½ inch." Sure enough, he discovered his aircraft had 7/16-inch valves and it was coming up on the time when it would be illegal to fly that engine without having them replaced.

The next weekend, one of those valves failed. The engine case broke open, oil spurted out over the windshield, and smoke filled the cabin. Klaus was able to land in a little field, just missing a power line. Quite rattled, he got out of the aircraft and walked toward a house over the hill to say, "I hope you don't mind, I just lost my engine and I landed in your field just over the hill there." Nobody had heard him come in because the engine had stopped, but he was greeted warmly and offered a drink of Scotch. "I had a very big belt," said Klaus.

Klaus sold that first Tri-Pacer and purchased another, with long range tanks. The following year, he decided to take the aircraft to attend a seminar in Saint Andrews, New Brunswick. "Just don't go over to Grand Manan," a local pilot warned him. "The airfield there isn't useable."

Naturally, Klaus had to check it out for himself.

The Seagull Takes Flight

Although the airfield at North Head on Grand Manan Island had been used by the Fundy Flying Club in the 1960s, and for outside charter companies to provide limited service to the immediate area, the field was too small. As Klaus saw it, "a new, larger airport was definitely needed for the island."

Intrigued by the setting and the challenges of Grand Manan, Klaus began to think about starting his own business. He had good childhood memories of helping in the family-owned resort operation and the freedom – and work – that came with such independence.

The chance visit to Grand Manan in 1978 led to a request for a transfer to the Fisheries Department, and ultimately a move to Grand Manan as Fisheries Officer. Trading his Tri-Pacer for a Cessna Model 172 Skyhawk, he then prepared to apply for a licence to provide a local air service in Grand

Manan. Transport Canada, however, acquainted him with their airfield requirements that prevented a charter service from being based out of the inadequate Grand Manan airfield. To Klaus, it didn't make sense. "Anyone from miles away could fly a charter to or from Grand Manan without any knowledge of the local, often treacherous, conditions," Klaus says ruefully, "but it *could not* be done by a local service."

Klaus found himself stuck between a rock and a hard place. The small population could not understand either why a local resident with an airplane wasn't allowed to fly them over to Black's Harbour just 20 miles north across the bay or to Digby 50 miles to the southeast, rather than force them to wait for the twice-daily two-hour ferry. Sure, without a commercial licence, Klaus could take someone over, but only as a non-paying passenger. This did not a business make. As usually happens, some flights were made, totally at Klaus's expense; at times he received "donations" to pay for gas. When asked to make emergency medevac flights, he couldn't say no.

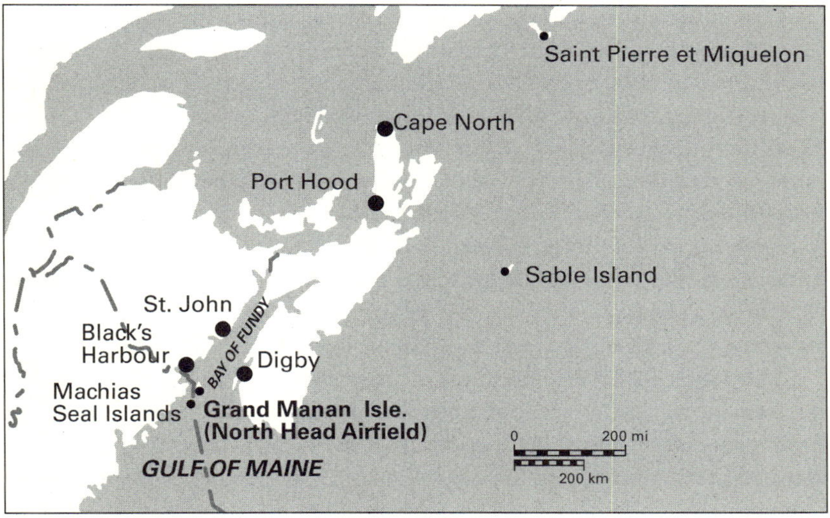

Klaus Sonnenberg's Aerial Neighbourhood

One strange day he was even approached by a member of the RCMP, who knew it was illegal for him to take passengers. Instead of trying to ensure that Klaus was following the rules, however, he requested that he break them. The officer was desperate: a fugitive had got onto the ferry and he'd missed it. Klaus thought this over carefully, thinking this might be a set-up. "I don't have a licence to take you," he replied cautiously.

"You have to! I've got to get over there before the boat lands. I'll pay you, no problem," the officer insisted.

"I'll do it on one condition," Klaus said. "You can pay me – I need the money for fuel – but *don't claim it*. Pay me out of your personal money, all right?" The deal was made. Klaus flew the officer over to collect his prisoner, and received $40 for gas. Then, the RCMP officer claimed the expense! A bit of trouble resulted from that, but in the end "good sense" won out and the incident was swept under the rug. "The rules were a bit artificial," Klaus comments.

In the summer of 1981 Klaus formally filed an application with the Air Transport Committee [ATC] of the Canadian Transport Commission, a regulatory body responsible for licensing and fare control in air transportation, for a Class 4 charter licence. Despite being supported by the entire voting population of the island, his application gained the not-so-positive attention of the enforcement branch. Klaus was charged with offering commercial air service without the benefit of a commercial air carrier licence as well as several violations for flying an un-airworthy aircraft. "Guilty!" Klaus would acknowledge, but adds "you have to know the circumstances."

His statement of defence declared that his Cessna 172 Skyhawk had been stripped in preparation for painting at Saint Stephen, New Brunswick, but just prior to the first coat being applied the painter took off to Newfoundland for the summer. Klaus had flown the aircraft back to Grand Manan so a local auto body shop could finish the job. They'd removed the wings for painting, and then Klaus re-attached them, but before he could have the work signed off by a certified Aviation Maintenance Engineer, and there were none on Grand Manan, a family emergency had come up – in British Columbia. Off he'd flown across the country. On his return to Saint John, he was charged. As Klaus explains it, "All flights that were made since the wings were reinstalled were considered a violation."

His pilot's licence had been already suspended for six months for operating an illegal air service, but he was unaware of this since he had not received the letter informing him of the suspension! Klaus says, "The minister had been asked to use his discretion and suspend me before the court date occurred. I found out at the court hearing." He appeared in court in January 1982 while his application for a charter still sat waiting approval with the ATC. He was fined $50, and the un-airworthy airplane charges were dropped for essentially technical reasons.

Manan Air Services

To aid in his application for a charter service, Klaus circumvented the "unacceptable-airfield problem" on Grand Manan by licensing Flagg Cove as his seaplane base, even though he didn't fly from the water that often. The company was registered as Manan Air Services.

Klaus Sonnenberg reflects on his ongoing attempts to establish a flying service on Grand Manan Island. "When we started, we had the little strip but after a year or two we realized that was a very big disadvantage. I purchased a Piper Apache twin up in Hamilton and flew it here, but the government wouldn't let me fly it out of this airstrip, even though they had done that in the 1960s. And, that plane was simply too old, but I did get some valuable twin experience on it."

Through Manan Air Services, the company also began to offer flying lessons under the name of "Atlantic Flyers". Finally, the name "Atlantic Charters" was chosen to cover everything from maintenance operations to flying.

Using an amphibious Lake Buccaneer, Atlantic Charters officially commenced business on Friday the 13th of August, 1982. Three months later, in November 1982, Klaus received the Air Transport Committee "A" Class 9-4, which allowed him to offer international charters as well as local flights. He was finally in business, legally. Well, mainly. "When we started flying out of here, there was no IFR [Instrument Flight Rules] approach," Klaus says. "We used to fly officially IFR when we went long distances just for safety reasons, but technically we were not certified as a company to conduct IFR flying because the base did not have an IFR approach."

Grand Manan had not only accepted Klaus Sonnenberg as one of their own, but so had a local lady named Melanie Ells, who'd been a passenger on flights as early as 1964 from the old Fundy Flying Club airstrip. Melanie, now married to Klaus and living at Deep Cove on the island, did everything possible to help make the Manan Air Services' businesses successful, while Klaus, the sole pilot, flew from dawn to dusk seven days a week. During the early 1980s, Klaus flew nearly 900 hours annually – mostly comprised of 30-minute trips – while avoiding as much as possible the worst flying months of March, April, and November, when storms almost prohibit Visual Flight Rules [VFR] flying and passenger flying.

The Lake Buccaneer aircraft had its limitations, so Klaus and Melanie negotiated permission to use the Cessna 172; and, because the old Apache had proved unsuitable, on July 18, 1987, they purchased their first twin-engine Piper Seneca. The family's steady income was still derived from Klaus's other work as general manager of the Grand Manan Fishermen's Association. Because it wasn't easy to make a living with the little airline while keeping their fees as reasonable as possible, he had to pick up other contracts.

Klaus Sonnenberg's knowledge and background in forestry and fisheries made him an ideal contract pilot for officials in pursuit of those who flaunt the laws of the land and sea. For the federal Fisheries branch, he patrolled Machias Seal Islands, an area disputed between the United States and Canada, photographing and using LORAN [Long Range Navigation] to document American scallopers fishing in sovereign Canadian waters. Animal predators were also on Klaus's radar. The Lands and Forests Department would hire him to go after night poachers in the Skyhawk "because you can see people in the woods at night so well," or to look for vehicles that had been reported lost or stolen.

Klaus Sonnenberg's continued efforts to establish his airline eventually gained him friends in high places, one of whom was Fred McCain, then Grand Manan's Member of Parliament. Through McCain, Klaus met with the Transport Minister of the day and got a commitment of ministerial discretionary spending of $1,000,000 for airport construction. With the additional help of New Brunswick MLA Eric Allaby, who was born on Grand Manan Island, an airport in the centre of the island officially opened in 1989.

It is an understatement that the air service was needed, especially for medical emergency flights. "Compared to earlier arrangements by ferry or local Coast Guard vessel, the 20-minute flight to Saint John has saved more than a few lives," Klaus notes with pride. "We continue to provide last minute flights as needed." The company now operates three all-weather multi-engine aircraft: a Piper PA-34 Seneca and two Piper PA-31 Navajos. As well, a single-engine Cessna 172 Skyhawk is used for sightseeing and local flights.

"When we first started we had no electricity, no pavement, and to fly in the wintertime was very challenging, to say the least," Klaus recalls. "In the summertime it was foggy, and no approaches, and even to this day it's a challenging area to provide air services for. Nowadays, of course, it's a bit different: we have instrument approaches here, and it's tamer."

However "tame" flying might have become from Grand Manan, Klaus still manages to work in some interesting and challenging flights, from Martinique in the Caribbean, to Lake Como, Italy, and to Canada's Arctic regions.

Flying for Fisheries

Klaus Sonnenberg's education and interest in biology, forestry and fisheries became a perfect fit for his maritime aviation enterprise. In short order his contribution to the fishing industry became invaluable.

"We have worked together with the Grand Manan Fisherman's Association ... [and] flown together to meetings around eastern Canada ... where he represented the fishers of Grand Manan and the Bay of Fundy," comments Dr. Shawn Robinson, a research scientist with the Department of Fisheries and Oceans at the Biological Station in St. Andrews, New Brunswick and an adjunct professor at the University of New Brunswick.

Dr. Robinson has great admiration for Klaus Sonnenberg's tenacity. "In all that time, I have seen him focused and ably representing his organisation. While some may not have always agreed with his position, you could not possibly deny that he came to the floor with knowledge, dedication and passion for his cause. I found that I appreciated those qualities in him and I think that the waters and creatures there in the Bay of Fundy have benefited from his efforts."

This was a time of dramatic change for the industry. New rules and regulations sometimes seemed to be in direct contrast to historical "rights", such as individual quota programs that provided fishermen with access to a specified quantity of fish.

On November 19, 1997, the Standing Committee on Fisheries was authorized by the Senate to examine and report on the questions of "Privatization and Quota Licensing in Canada's Fisheries". Various opinions were sought and quoted in a portion of the May 28, 1998 report called, fittingly enough "The Controversy". Michael O'Connor, Fleet Operations and Government Regulations, National Sea Products, stated: "The Minister of Fisheries and Oceans has discretionary licensing powers and can revoke or change our licence conditions. The latter point creates uncertainty about yearly sector shares and quotas, and can destabilize investment plans, marketing and production plans."

The committee learned that with fish stocks undergoing unpredictable fluctuations, catch limits often resulted in a "race for fish" at the beginning of the new season from fear that fishing would be halted prematurely. This and other counter-productive practices had seriously undermined communal spirit among the fishermen.

In Canada, initial allocations within a fleet are based on the fisher's "catch histories", the amount of fish caught in previous years, with the aim of rewarding those who have been active and have demonstrated an attachment to the fishery. But these records may be flawed for various reasons.

Klaus Sonnenberg, speaking as treasurer of the Eastern Fishermen's Association, noted, "Every fisherman realizes that, sooner or later, there will be an allocation based on past performance. The result is that people who have more than one licence, instead of fishing each one in turn, are fishing all of them, if possible, with different vessels, thereby creating what is called 'history' so as not to be discriminated against."

Klaus further stated: "What we are finding, for example, in the lobster fishery, is that too often people are fishing increasingly harder simply because, if it is turned into an IQ [Individual Quota] fishery, they want to make sure that their history is such that they will become millionaires in the future by the present use-it-or-lose-it policy that Minister Tobin condemned."*

Fishermen were encouraged to discard low-value or undersized fish, which counted against their individual quotas, for higher-value fish. This destructive and grossly wasteful practice, known as "high-grading", had unfortunately become accepted as the way of doing business in the Atlantic fishery. The committee was made aware of this "dumping and discarding" practice during the mid-1980s. Finally one fisherman, so disgusted by this practice, had written to Fred J. Mifflin, Minister of Fisheries and Oceans in 1996-97, to describe having to "plough his boat through dead fish for miles around."

Among a number of recommendations, the Senate Standing Committee on Fisheries suggested that the issue of privatization and individual quota licensing in Canada's fisheries be debated in the Parliament of Canada, and that no new individual quota licences be issued in Canada until written public statements on individual quotas and partnerships agreements were issued and a debate had taken place.

Trouble had also been brewing with a cited lack of understanding between the Canadian Council of Professional Fish Harvesters and the Eastern Fishermen's Federation, comprised of 21 Atlantic Canada fishermen's organizations from New Brunswick to Quebec. Piqued by the national organization's demand for professional standards and training assessments, the Maritimes group "jumped ship". The Grand Manan Fisherman's Association was one of the 21 dissidents, and both Melanie and Klaus Sonnenberg were at the helm, at various times being members of the federation board and local council. The problems were multi-faceted, from the training requirements for fishers set down by the federation, to the increasing jeopardy to the wild fisheries from aquaculture (fish farming), to a "lobster war" with American fishers.

As director of the Grand Manan Fishermen's Association, Klaus acknowledged there was room around the island for aquaculture, which had come to employ a large percentage of the local population, but added, "That room's been taken up. Every salmon site that goes in now displaces a significant portion of traditional fisheries."

In spite of fierce protest and a blockade of one harbour on Grand Manan Island, five new salmon leases were approved in June 2001, with one of the sites forcing the closure of an active herring weir. Klaus spoke

* Brian Tobin, Minister of Fisheries and Oceans from 1993-96, had mounted a fierce campaign against foreign over-fishing of waters on the edges of the Grand Banks, located just outside Canada's declared 200-nautical-mile Exclusive Economic Zone [EEZ]. In April 1995, Tobin's department was embroiled in the so-called "Turbot War", which pitted Canada against the European Union. For his roles as Fisheries Minister and in the referendum in that year, he earned the nickname "Captain Canada".

up for others' rights, as he'd so often done to defend his own: "They shut down a weir that's been catching fish almost every year for 80 years! ... Proven fisheries should not have to suffer for the sake of an emerging industry." He envisioned the disappearance of a way of life, with fewer and fewer family-owned operations in existence, and local aquaculture leases being controlled by three multi-national corporations. "The people here will end up as servants to a corporation with its headquarters in another country," Klaus laments.

Klaus and Melanie Sonnenberg's defence on behalf of local fishers is an ongoing struggle.

"New Brunswick lobster fishermen declare open season in disputed waters," stated CBC News on August 15, 2002. The lobster fishing area around Grand Manan Island is called District 38, but this problem focused on District 38B, a 110-mile "grey zone" between Grand Manan and Maine to which both Canada and the United States claim ownership. The Canadian lobster season usually starts the second Tuesday in November and ends on June 30. Americans can fish for lobster any time during the year. While Grand Manan fishers usually set out about 1,500 traps, the lobster fishers from Maine set between 15,000 and 25,000 traps in their year-round season.

Angered by the incompatibility in dates and numbers with their American competitors for the lucrative lobster fishing harvest, and supported by the federal Department of Fisheries, in August 2002 – a full three months ahead of their traditional schedule – a number of Canadian lobster boats set out their traps. Melanie was then Project Manager for the Grand Manan Fishermen's Association, while Klaus held the position of General Manager. Klaus was quoted on the news broadcast, saying, "You can see where the line is in September, October, because that's where all the American traps are, waiting for Canadian lobster fishers to cross that line," adding that the Americans know that lobster are more plentiful in Canadian waters.

The following year, on September 26, 2003, Fox News Network announced that the problem had escalated to overcrowding, which resulted in tangled lines, flared tempers and lost equipment.

In April 2004, a Canadian/U.S. Lobstermen's town meeting was held in Portland, Maine, hosted by the Lobster Institute. Klaus sat on the planning committee, representing Grand Manan. Over 165 concerned people attended the meeting, coming from as far south as Florida and north from Nova Scotia, to discuss the status of the lobster resources with respect to health and nutrition, water quality, and the need for consistent, long-term monitoring. Attendees were given an opportunity to share observations from their day-to-day operations. Few could have been more informed than Klaus Sonnenberg, who over the years had viewed the situation from the air

– the numbers and positions of fishing boats – as well as from his long-term background involvement with the fishermen's associations.

By 2005 the economic situation for fishers had worsened to the point that Connors Brothers' sardine processing plant on Grand Manan had closed, putting 150 islanders out of work. Klaus, in his continued position of General Manager of the Grand Manan Fishermen's Association, was quoted in an article in National Fisherman (March 2005) titled "As processors close, a way of life goes by the board", saying that while the economic blow was significant, the impact of closure of the nearly century-old plant went even deeper. "Whole families worked there, some for several generations. This is part of the life in Grand Manan."

And then the whales moved in to the Bay of Fundy, and stayed. The rare and endangered North Atlantic right whales – hunted to near extinction in the late 1800s and now a protected species, that marine biologists estimate to number at no more than 350 – normally vacate the bay by mid-October. They then move quickly into the Gulf of Maine before heading farther south. However, in 2006 they apparently found the food supply to be so plentiful that they'd stayed longer, which was seen to interfere with the traditional start to the Canadian lobster fishing season. Officials from the Department of Fisheries and Oceans therefore threatened to delay the opening of lobster season in fear that the whales would become entangled in fishing gear.

With more than 1,000 fishing families depending on the lucrative Bay of Fundy lobster fishery for their income, in November 2006 the Canadian Press sent reporters to inquire about the seriousness of the situation. Klaus Sonnenberg, on behalf of the Grand Manan Fishermen's Association, expressed his view that the recently appointed managers of the Fisheries Department were overreacting. "I'm a pilot and I've been flying the Bay of Fundy on a daily basis since 1978," he said. "It's normal to see right whales lingering in the bay at this time of the year. We've never had any problem." He added that U.S. fishermen fish for lobster year-round in the Gulf of Maine, and there had never been a need to close the fishery to guarantee the right whales' safe passage.

And on it goes. In 2009, the catch was up but prices were down, and Gulf Coast fishermen held a major protest in June to highlight their plight. The advantage of Grand Manan fishermen, said Klaus, was that they fished for more than just lobster. So somewhere, something must be holding its value.

Just as Klaus Sonnenberg had picked up the aviation bug through his father's stories of flying in Germany, he has shared his passion for aviation with his wife and sons, who have also taken up flying. Whether soaring with the seagulls or working to pacify the politics surrounding the volatile fishing industry, Klaus Sonnenberg and family continue to relish the challenges presented to them in their chosen home, the beautiful and exotic island of Grand Manan.

Hijacking over Prairie Skies

Many hijackers have attempted, with poor results, to copy D.B. Cooper's feat of November 27, 1971 of jumping out of an aircraft in flight with a parachute and a bag of ransom money.* But Cooper himself may have taken a page from someone else's crime book.

On Friday, November 12, 1971, just two weeks before Cooper's famous caper, a swarthy, dark-haired man purchased a flight ticket in Calgary under the name of Dennis Munro. He calmly boarded a McDonnell Douglas stretch DC-8 on Air Canada's Flight 108 departing from Calgary International Airport en route to Toronto International Airport with a crew of nine and 118 passengers.

While Munro's actions brought odd looks when he insisted on wearing his long black coat while other passengers stowed their wraps in the overhead bins, soon all were buckled into their seats and the flight departed on schedule.

Shortly after takeoff, approximately the time the plane was crossing into Saskatchewan airspace, Munro got up, still wearing his heavy overcoat, and went to the lavatory. He emerged with a new look.

He darted in to the galley wearing a mask over his head and waving an 18-inch-long sawed-off double-barrelled shotgun. "Over here!" he barked

---

* On November 27, 1971, a man who became known as D.B. Cooper hijacked a Northwest Orient Boeing 727 airliner in Oregon. His demands for money and parachutes were met at Seattle-Tacoma airport. Then, the hijacker jumped – into history and "folk hero" status. His identity remains a mystery, and only a small portion of the money was ever found, in a water-soaked bundle on the north bank of the Columbia River, near Vancouver, Washington. See Chapter 9, "The Cooper Vane and How it Came To Be".

to the purser, John Arpin, who was at the bar mixing a drink for one of the passengers.

Arpin turned to stare down the barrels of the gun. "You're kidding."

"I'll blow your head off! I'll show you if I'm kidding. Sit down!"

Arpin sat, and the bandit shoved a note into his hand. "Take this to the flight deck. And don't read it or I'll blow your head right off!"

Turning to flight attendant Mary Dohey, he pointed the gun toward a seat. "Sit beside me!"

Keeping Dohey at gunpoint, Munro carefully set two small bundles on the first class lounge table. Each contained five sticks of dynamite. "Are you nervous?" he asked.

"Yes, a little bit."

"I'm sorry, lady, to do this to you," the hijacker responded, solicitously taking her hand.

John Arpin had just left the flight deck after delivering the unread note to the captain when he heard an explosion. The bandit had blown the flight attendant's head off and he'd be next!

The gun had indeed gone off – whether intentionally or accidentally – blasting a four-inch-diameter hole in the plastic sheeting of the partition that separated the first-class area and the small kitchenette-bar.

The purser quickly hurried out to assure passengers in economy class that a light bulb had exploded. As they had not seen the masked man emerge from the washroom, they remained calm.

But this incident further agitated the bandit, who then "went quite berserk," in Arpin's words. When he asked for a cigarette and the purser picked up someone else's pack, which was not his brand, he accused the purser of trying to dope him.

"Your heads are gone," he warned.

Then he smiled, and flicked his butane lighter against the fuses clutched between his fingers, which led to the dynamite. The flight attendant and purser watched in horror as the four-inch flame licked the ends of the fuses. Then the lighter was withdrawn – only to be teasingly applied once more, just one-quarter inch away, then brought back.

John Arpin tried to keep his voice calm. "I'm sure you're a very nice person," he assured the hijacker, "and that you don't plan to hurt anyone. We'll do exactly what you say. It's a company policy to do nothing to antagonize a hijacker. Our first concern is the safety of our passengers."

Mary Dohey sat still, drawing upon her experience as a nurse to overcome her terror and so prevent the man from becoming even more agitated. "Most of the time I held his hand in my right hand, and the wires leading to the fuses in my left hand," she later stated to *The Albertan* reporter Stephani Keer.

As Munro waited for the captain's response to his note, he informed Arpin and Dohey that, in addition to the 10 sticks of dynamite in front of them on the table, he had 60 more sticks hidden in a parcel under the seat. Indeed, two wires held in his gloved right hand that ran from his coat to the space beneath the seat seemed to support his statement.

The note, duly delivered unread by the purser, commanded Captain Vern Ehman, First Officer Nels Hagenson, and Second Officer Noel Bellanger to turn south to Great Falls, Montana. Then, the captain was to order Air Canada officials to arrange for $1.5 million in American currency to be ready upon landing and delivered to the aircraft in Great Falls by a woman. After receiving the money, the aircraft was to fly to Regina so the other 117 passengers could disembark, along with their baggage. Then the aircraft was to take him, Munro, over the Polar Route to Ireland. "I sympathize with the efforts of the IRA [Irish Republican Army], and I'm giving all the money to them," the hijacker told Mary Dohey.

The next to come was a bizarre list of instructions that had the airliner making aerial leapfrogs between Canadian and American airports while Munro plotted his next move.

The captain immediately radioed ahead to Great Falls, informing them of the situation and adding that he felt the hijacker was extremely serious, and dangerous. Therefore, no police intervention should be present at the airport to further agitate the already volatile man. Captain Ehman was then forced to circle over Great Falls for 90 minutes until informed that the money was ready at the airport.

Meanwhile, John Arpin kept assuring Munro that Air Canada was doing everything possible to meet his demands, while Mary Dohey continued to hold his hand. Both talked to him in low voices about everything from their respective families to attendance at the Calgary Stampede.

On the ground, Great Falls authorities, working with local businessmen, hustled to bundle as much money as could be found in such a short time – in fact only $50,000 – into a briefcase and make it ready for delivery. The money was brought to the airport by a female police department office worker and the briefcase was hoisted up by a rope lowered from the cockpit. Giving it a cursory glance, apparently believing that it contained the full $1.5 million, Munro ordered the captain to take off for Regina.

But the Regina airport had closed because of weather, and this information was relayed to the hijacker. "Okay. Another airport then, close by." How about the military base at Moose Jaw? "Fine." Flight changes were made, and covertly the military police in Moose Jaw were put on full alert.

Then, another surprise. While en route to Moose Jaw, the hijacker ordered the aircraft to return to Great Falls. Flight plans were amended once again, and back they flew. In Great Falls, Munro allowed the passengers to disembark "slowly and cautiously, one row at a time." Their baggage was also permitted to be offloaded. Munro ordered that his own suitcase be brought to him from the cargo hold.

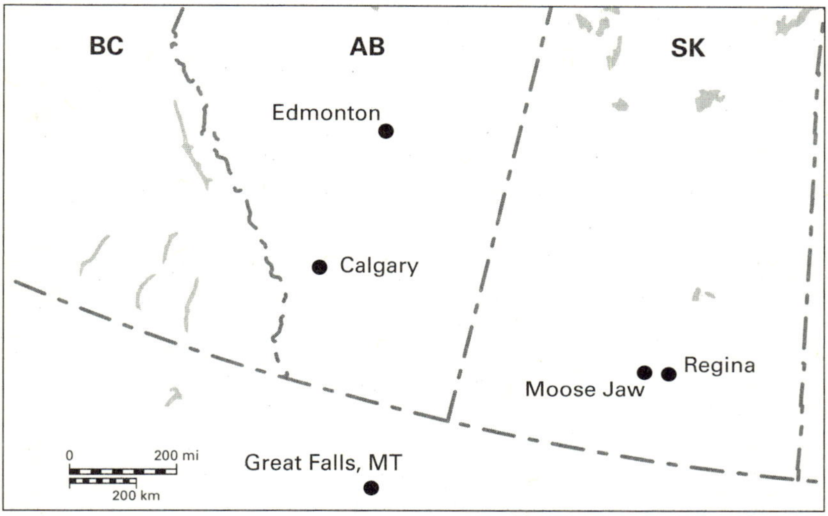

Flight 108 is Hijacked

The aircraft needed refuelling, and he nodded, yes, that could be done, but from a runway far away from the terminal.

Even though the landings and takeoffs must have seemed quite unusual, this was the first time passengers learned that their aircraft had been hijacked. Still dazed from the realization, they were ordered to go through customs as usual, and told they would be picked up later that night by a relief plane and flown back to Canada.

Inside the DC-8, sitting for the second time this night on the tarmac at the Great Falls airport, the bandit became extremely nervous. "The FBI are going to shoot me!" he cried, and suddenly ordered the refuelling to stop. "I don't care if the tanks are only half full. Take off again! I'll tell you which direction to go when we're in the air."

The captain duly followed the orders, and the DC-8 became airborne once more.

"Go to Phoenix!" Munro commanded.

"Okay," replied the captain. "We have just enough fuel to make that destination."

Then, "No, turn around, go to Calgary."

The aircraft circled and swung back over the U.S./Canadian border, for the fourth time on this fearful flight, at 11:00 pm. Unknown to the hijacker, Malmstrom Air Force base at Great Falls was providing two fighter jets to escort the DC-8 to its destination in Calgary – or wherever the hijacker might decide to order it to head next.

"I want to talk to the captain!" Munro suddenly announced. "Bring him back here."

With the safety of his crew and the aircraft at stake, Captain Ehman turned the controls over to the First Officer, informing him and the Second Officer that he might have to resort to drastic action. He then went back to the First Class cabin to meet the hijacker face to face.

Munro greeted him with a new demand. "I want a new crew to take over in Calgary. And we're going to pick up another passenger when we land."

But, before the captain could absorb the ramifications of this request, a new, urgent demand was made: "Before we approach Calgary I want you to lower the aircraft to 3,000 feet. I'm bailing out."

"We were losing altitude then," the captain recalled. "I wanted to get down early to depressurize the plane. He told me to slow the plane to 100 mph. I said it was impossible but we can give you 110. That's when he said he was going to jump. He said that he had a parachute in a parcel, and he wanted to go out the main door."

"It is impossible to open the main door because of the jet stream," Captain Ehman informed him, "but you *could* go out an emergency exit over the wing, and slide along the wing."

The hijacker shook his head. "I don't believe you." Then he menacingly indicated the 10 sticks of dynamite still sitting on the lounge table. "I'll blow the end out of this plane!"

He and the captain stared, eye to eye. Then Munro said, "Give me a knife, right now. I'm cutting this parcel open and putting on my parachute."

The Other Side of the Saga

Meanwhile, an emergency command post had been set up in Calgary in the terminal control tower (the present location of the Shell Aero Centre). When informed that the aircraft would be landing for fuel, the command post group began devising a plan: Was there a safe way to board the DC-8? And who would be the first persons to do so?

On being informed of the hijacking, Calgary Airport Manager Bill Watts and Airport Customer Service Manager John Robulak had raced from their homes to the airport. There was not enough time to organize police escorts for them, so the managers barked out descriptions of their vehicles over the phone and hit their gas pedals, shattering every speed limit getting to the site.

There, Robulak went directly to the Air Canada centre to monitor communications. Because of limitations of range it was first necessary to set up a complicated relay, from the flight itself to airports in Regina, Vancouver, and then Calgary. Once the connection was in place, Robulak heard the captain's warning: "The hijacker is desperate. He must be taken seriously!"

THE ALBERTAN

Price 15 Cents — Calgary, Alberta, Saturday, November 13, 1971 — Phone 263-7730

Calgary flight hijacked

STRETCHED DC-8 WAS ORIGINALLY BOUND FOR TORONTO

From staff and Canadian Press

The hijacking of an Air Canada DC-8 and subsequent hours of aerial leapfrog between Can[...] and Montana cities ended here just after 12 a.m[...] morning with the delivery of the overpowered hij[...] at Calgary International Airport.

The hijacker, who had left 118 fellow passe[...] on Flight 812 stranded in Great Falls, Montana o[...] second of two stops Friday, was overpowered b[...] plane's captain some time after taking off from [...]tana at 11 p.m.

Air Canada officials later said the hijacker threate[...] blow out the back of the plane unless crew opened an [...]gency window. As he bent over to put on a parachute, [...] member hit him over the head with a fire axe.

The aircraft just made it before the airport was [...] down by fog. All incoming flights were being diverted by [...] today.

There was apparently a small explosion on board th[...]craft on landing or just prior to touchdown. The aircra[...] under exteremely tight security and it is expected it [...] gone over minutely by security and maintenance offic[...] sure it did not suffer any damage.

The hijacking created a midnight traffic jam at C[...] the international airport.

Police were diverting hundreds of cars after radio s[...] announced the plane would land at 11:45 p.m.

Members of the media were admitted only after ide[...]tion was closely examined. Passengers on out-going [...] were required to produce tickets before being admitte[...] roads leading into the airport.

During the six hours the passengers were held, the [...] — originally on a Calgary-to-Toronto route — was dive[...] Great Falls, took off again for Regina, then mysteriou[...] turned a second time to the Montana centre when t[...] aboard were released.

The hijacker, who said he was a member of the Ir[...]publican Army and willing to die for his cause, was r[...] armed with six sticks of dynamite and a gun.

One shot was fired in flight and an Air Canada spok[...] said there was some indication the fuselage of the pla[...] been punctured.

The large jet was escorted to Great Falls by United [...] Air Force fighter planes.

After being commandeered about 4:15 p.m. MST, th[...] had circled Great Falls for about two hours while $50,00[...] sum demanded by the hijacker was being raised.

The DC-8 had eight hours of fuel when it left C[...] where 101 of the passengers boarded. The fuel it took on [...] the first stop in Montana lasted about half an hour.

The gunman had asked that a rope be lowered fr[...] cockpit to enable the ransom to be brought aboard.

An FBI agent talked briefly with the captain, Vern [...] of Montreal, but was told the hijacker had a radio. The [...]sation then ceased.

The $50,000 was raised by Great Falls bankers a[...] delivered to the airport by a woman whom officials sa[...] not been connected with the hijacking. It was carried in [...] case.

The hijacker's plan had been to leave the passengers [...]gina, obtain a fresh crew and fly directly to Ireland.

An Air Canada spokesman said there were no crews [...]gina qualified to fly the overseas jet. However, a navig[...] quired on transatlantic flights, had been flown in from [...]peg.

The atmosphere on the tarmac at Calgary was t[...] RCMP and city police cars, two ambulances and six fir[...]

The smiles are enough to describe the relief felt by more than 100 passengers released late Friday night in Great Falls, Mont., from an Air Canada [...] The long ordeal included a landing, a take-off and another landing at Great Falls where the masked unidentified hijacker let them go.

The Albertan, *November 11, 1971*

A national radio network quickly connected with Vancouver operations to improve communications. Other radio operators were then called in to the Calgary airport to monitor all calls, and maintain contact with the RCMP, city police and fire departments, as well as with Western Airlines in Great Falls, who had received the DC-8's passengers after they'd disembarked and gone through customs.

"Make the passengers as comfortable as possible," Robulak requested of Western Airlines' manager. "Open the bar and give them any refreshments they might need."

However, it was found that the Western Airlines aircraft could carry only 94 of the 117 passengers, so the crew of an incoming Air Canada flight that had just arrived from Toronto was asked to stay on alert in case they might be required to go on to Great Falls to pick up the remaining passengers. The Regina Airport also assembled a fleet of buses to meet passengers from the pick-up flight departing Great Falls and transport them to their various destinations, or to the closest airport to take them on to Toronto, the original destination of the hijacked flight.

On the chance that Munro was serious that the flight would next have to take him to Ireland, Captain Les Anderson, based in Winnipeg, was making arrangements to fly a navigator to Calgary to accompany the hijacked aircraft overseas.

Then came the hijacker's next request: "I want a ground power unit to provide electricity to the plane while the engines are shut down – and an air-starter to start the engines again."

Back at the Calgary International Airport, John Robulak and Bill Watts finalized plans for the next stage, when the DC-8 touched down, due at 12:45 am. Although the hijacker hadn't mentioned fuel, an oil company was alerted in case it might be needed.

The next move would be crucial: Bill Watts, John Robulak and members of the airport's operations and ground crew met, along with the RCMP and City of Calgary police, fire and ambulance units. Watts arranged for a radio car to carry three volunteers, the minimum staff for servicing the aircraft, to go out and meet the DC-8 when it landed. One person would operate the requested ground power unit, another the air-starter, and the third remain in the car to report the action to the control tower.

On the streets of Calgary, a radio station announcement that the hijacked DC-8 was soon scheduled to land had caused a major traffic jam, obliging police to re-route hundreds of vehicles. Reporters and other members of the media were admitted past barricades only after showing identification, and passengers scheduled for outgoing flights had to produce their tickets before being allowed onto roads leading to the airport.

To further add to the confusion and congestion at the Calgary airport, Edmonton's airport had closed due to fog, and two flights had been diverted to Calgary. But the hijacked aircraft was given priority for all services, with other incoming flights delayed or diverted if at all possible.

Then came the next order from the DC-8. "We need an ambulance and a demolition squad!" What was going on inside that aircraft?

Back inside the Airliner

When Munro asked the pilot for a knife to cut the twine holding the parachute parcel together, Captain Ehman coolly replied, "I don't have a knife but I have a fire axe." Somehow the captain had the idea that there was no parachute in the parcel, but the hijacker's mental state was so unpredictable that it was best to go along with his requests, and keep him as calm as possible.

The bandit, still holding the sawed-off shotgun, handed the parachute package to the captain and grabbed hold of the fire axe. Looking confused, he then put the gun down on the table to cut the twine with the axe. The tall, wiry captain saw his chance.

Grabbing the man's gun, he flung it down the aisle and grabbed the bandit around the throat with both hands. The purser lunged from the rear of the first class cabin to jump on the wildly thrashing man. Although Dennis Munro was small-built, five feet eight inches in height and weighing 120-130 pounds, in his agitated state he fought like a lion.

A roll of duct tape was produced, and both the captain and purser attempted to tape him up, but he got one arm free. "I'll blow you all up!" Munro cried, as he tried to shove his hand inside his pocket.

Did he have a bomb? Very possible, judging from his constant threats to blow up the airplane! Best to take no chances! Purser John Arpin snatched at Munro's hand just as he was reaching inside his pocket.

Then, grabbing the axe, Assistant Purser Bonny hit him a crack with the axe handle, temporarily knocking him down. The hijacker's hands were quickly taped together, but on he fought, thrashing about on the cabin floor. Two more whacks on the head did little to subdue him. "Use the sharp side, for Christ's sake, use the sharp side!" someone yelled.

They nailed him – with the blunt end of the axe – and finally knocked him out. The flight crew then taped him securely to a seat, as the captain hurried to the cockpit to help complete the landing.

"The hijacker has been secured!" First Officer Hagenson radioed to the ground contacts. Whew! Six fire trucks and two ambulances then escorted the DC-8 to a berth at the main terminal building on the west side of the airport when it touched down at 11:59 pm. Portable passenger steps were wheeled out for an RCMP inspector and Airport Manager Bill Watts to enter the aircraft.

An eerie calm had settled inside the DC-8, with the bandit secured and appearing to be out cold. The police inspector went up to the man and gave him a pinch on the arm. The hijacker jumped, and opened one eye. "He's not unconscious," the officer said with a hint of a smile. "He's faking it. Keep him secured."

Knowing that 118 passengers, a nine-member crew and a $10 million aircraft had been at great risk, the desperate situation had called for calm and effective responses. The captain and all crewmembers were gratefully congratulated for the professional manner in which they'd acted to bring the affair to a safe conclusion.

"He was completely deranged, and bent on self-destruction," Captain Ehman said later. "I felt he would have blown us all up before he jumped."

And so the saga came to a close at Calgary International Airport, and none too soon. By 12:30 am all incoming flights had to be diverted to other airports because of an intense fog that had blanketed the area.

Fortunately, no one had been hurt in this incident, except for the hijacker, and then only superficially.

Who was this Masked Man?

On an aircraft, one never knows what goes through the minds of one's fellow passengers, when strangers are crowded together on a flight with no escape except through the normal routine of disembarkation, hopefully at one's point of destination.

The hijacker had certainly planned ahead for this caper. Police recovered the sawed-off double-barrelled shotgun, two live rounds of ammunition, and parts of an expended round that had been fired into the partition separating the flight deck from the first-class cabin. Although he'd said he had 70 sticks of dynamite, police reported that somewhat less than that amount was found inside the airplane. And no parachute was found inside his wrapped parcel.

The true identity of Dennis Munro was established as Paul Joseph Cini, age 27, through fingerprints taken at the General Hospital in Calgary where the hijacker had been transported. Contact with the fire-axe had left him with a concussion and a two-inch gash to his head, but otherwise he seemed in satisfactory condition.

As is often the case, Paul Cini was described by his neighbours in Calgary, where he lived with his parents and younger brother in a modest duplex on Arlington Drive SE, as "quiet and aloof." The family, who generally "kept to themselves," had moved to Calgary from the United States some time earlier, and five years before the hijacking incident Paul had come to join them.

Cini had been employed as a delivery truck driver, but rumours circulated that he had been a U.S. Army paratrooper. In the months prior to the hijacking incident, neighbours had seen him hanging around the house during the hours he had formerly been working, so it was thought that he'd recently become unemployed. He was a backyard mechanic, and drove a number of different vehicles during his five-year tenancy at Arlington Drive. He was also engaged to be married. Had dwindling finances driven him to such a wild and desperate plot? This is one of the many unknown facts, along with the question of why some aspects of his plan had been so confused and poorly executed.

When a representative of the Irish Republican Army was contacted while on a cross-Canada fundraising tour, he dismissed any connection with Paul Cini or his bizarre hijacking plan, calling him "some nut who had nothing whatsoever to do with the IRA."

Paul Cini was considered too ill and unstable to appear in provincial court the following Saturday to answer charges of "interfering with transportation facilities," but he eventually was tried and convicted, and received a jail sentence.

Follow-up prevention of such incidents in future resulted in the introduction of metal-detection devices at Calgary International Airport similar to those already in place at Vancouver, Montreal and Toronto airports. Up to that time, Calgary had been relying on a detector that switched to various loading gates, but was not always in use.

As well, legislation was enacted by then federal Justice Minister John Turner to deal with hijacking on an international front, with the adoption of both The Hague Convention for the suppression of unlawful seizure of aircraft and a further convention dealing with sabotage. Amendments to the Criminal Code were also endorsed. These moves were made following severe criticism by the Canadian Airline Pilots Association for the government's failure to take legal and security measures to prevent sky piracy.

And so, the Munro/Cini affair came to an end – so far the only time such an incident has occurred over prairie skies.

When Your Time Has Come ...

One of Canada's pioneers of aviation, James Joseph Collins was a Newfoundlander through and through. Born in St. John's in 1925, his introduction to aviation came through the exciting wartime adventures of his older brother Bill, who was a radio operator ferrying aircraft from Gander overseas.

Then came tragedy. On March 13, 1943, just three days after Jim's 18th birthday, Bill Collins was in a PBY Catalina flying boat returning to Bermuda from an aborted flight to the United Kingdom. An ill-judged attempt to make a "glassy" landing on still water in the setting sun, which distorted altitude perception, resulted in a fatal crash. Of the six-man crew, only the co-pilot and flight engineer survived. For Bill Collins and three other crewmembers, including their captain, "Duke" C.A. Schiller, one of Canada's greatest bush pilots, their time had come, suddenly and tragically.

Notwithstanding the terrible accident that had ended his brother's life, Jim Collins decided that aviation would be his chosen field. Having started out as an auto mechanic, he switched careers in the late 1950s. After gaining an aircraft maintenance engineer's licence, followed by a commercial pilot's licence, he founded an aircraft charter service, Air Transit Limited.

Because roads to rural coastal communities at that time were almost non-existent, the sea had traditionally served as the "highway" for many Newfoundlanders. Jim's aviation company planned to provide a faster, and much-needed, service to connect fish company owners with plants and suppliers along the coastline of Newfoundland and Labrador. Gus Etchegary, an executive with Fishery Products Ltd., quickly became a frequent flyer with

Air Transit Limited. Like Jim, Mr. Etchegary was a Newfoundlander born and bred, and was convinced that Newfoundland's joining Confederation had led to the mismanagement of what was then its only resource – its fisheries.

Jim Collins, like any pilot who flew the coastal areas, learned to constantly watch the weather. Bush flying anywhere is a difficult and dangerous way to make a living, but in Newfoundland the weather conditions could be different on three sides of the island at the same time. The best, and sometimes the only, time for safe bush and float flying throughout Newfoundland occurred between May and October, with the winter season being occupied by maintenance work.

To survive financially, the company also flew medical evacuations, sportsmen hunting and fishing trips, mining surveys, and the odd charter trip to places such as Saint Pierre and Miquelon. Jim also operated an aircraft on wheels that he could use when needed to bring management back and forth from Bell Island iron ore mines to St. John's.

A Family Affair

Jim Collins's home in St. John's became the company's dispatch office, with the transmitter and receiver ("transceiver") located in the kitchen. Jim's wife, Eileen, became the licensed operator and handled everything from booking flights to issuing invoices, along with caring for their six children (Jeff was born in 1955, Paul in 1957, Brenda in 1958, Jane in 1959, and the twins Dave and Daphne in 1966). The aviation business gave the Collins kids an unusual childhood. Brenda and Jeff recall just how different they were from neighbouring families. "Being around the transmitter all the time meant we grew accustomed to the call letters and cryptic radio talk," Brenda says. "It was not unusual to hear one of us kids, while playing, say, 'Roger' or 'Over and Out.'"

Nearby Octagon Pond (later renamed Paddy's Pond) was ideal for a seaplane base – large enough to allow safe takeoffs and landings, and close to St. John's to give easy access to passengers. On weekends, Eileen would bundle the kids into the station wagon, along with a picnic lunch, and wait for Jim to fly in while the children played in the water or on the dock. Sometimes he'd take them for a short spin in the aircraft, their version of a Sunday drive. "Before I was able to touch the pedals I would hang onto the square wheel in the Cessna, thinking I was co-piloting the plane," Brenda recalls.

"Dad was more inclined to take the boys up than the girls, it was an old-style society," she adds, "but he did take us up on occasion. He often flew Jeff and Paul to Middle Ridge, where a group of friends had a hunting lodge."

Watermelon and Yogurt

At one point, Jim had a contract to fuel aircraft that were flying supplies from the United States to American troops based in Reykjavik, Iceland. One day a big transport aircraft came in, loaded with yogurt and watermelons.

"That was all they were carrying," Jeff says, "but they were carrying enough yogurt and watermelons for an entire American base! When the plane came in it had a mechanical problem, and could not go on. Well, there wasn't enough refrigeration in all St. John's at that point to put up all this stuff. The pilots basically said, 'We can't deliver it, so we'll have to get a truck and haul it to a landfill because it's going to start to go bad.'

"Do you realize how *rare* it was to see watermelon, out of season, in Newfoundland?" Jeff asks. "And yogurt was never on the menu. So the calls went out. The airport crew distributed this stuff all over town, to people's homes, to the Salvation Army, to all the food-gathering places. There were long lines of people waiting for yogurt and watermelon."

"It was the first, and only, time that we ever saw that," Brenda adds, recalling the event.

It was most residents' first taste of yogurt.

"I remember being fascinated with the stuff; we'd simply never had it before," Jeff says. "And the watermelon! When we did have it, one slice would be taken out, then Mom would lay it down and cut it into six little wedges, one for each of the kids, and everybody would have a little wedge. Now here we were, able to have as much watermelon as we wanted. We even made necklaces out of the seeds."

"And Mom made pickles out of the rind," Brenda adds. "Those awful, awful pickles!"

Adventures in Flying

Jim took his oldest son Jeff for his first airplane ride when he was only four or five years old. When they returned home and Eileen found out he'd been taken up in an old cobbled-together Stinson that was little more than a hobby aircraft, she really chewed him out. "I guess it was not a very good aircraft," Jeff recalls. "In the video footage we have, taken out the window of this aircraft and panning past the instrument panel and struts, you could tell that it was kind of bare-bones. It may have been a salvage job, some Stinson that went down and they'd recovered. It seemed to be one of those jinxed planes. It had a couple of accidents while they had it, and on its way to being sold it crashed nose down into a bog. It was salvaged but didn't last beyond the new owner."

When Jeff was 10 years old, Jim took him along again on a flight to Catalina, a small settlement on the east coast of Newfoundland. Jeff was excited to be riding in the yellow and black Cessna 172 aircraft, but it would not be a routine trip. They flew in on wheels, to land with a thump on the shale and gravel runway, where a sharp stone punctured the nose-wheel tire. When they were ready to leave, Jim got help to pump it up. Their takeoff was successful, but then there was the landing to consider. Jim radioed ahead to St. John's airport to warn, "I'm landing with a slack nose wheel tire." He then patiently explained to young Jeff that he would try to land on the back landing gear and keep the nose up as long as he possibly could.

The control tower radioed back, requesting that he do a fly-by so they could take a look at the nose wheel to ensure that he hadn't bent the gear in the takeoff. They confirmed that the gear seemed to be straight enough but the landing would be tricky. To be safe, they would scramble the fire trucks. This is what Jeff saw lined up at the end of the asphalt airport runway. "Dad, they've got all the fire trucks out! Look at that!"

"Yes," his dad replied soberly, "they're for us." He then gave Jeff a warning. "Look, this may go badly. If we do flip, the seat belt will keep you in the seat. Wait until the aircraft stops, open your seat belt, and *get out*. Don't worry about me; don't worry about anything except getting out of the aircraft as fast as possible. They'll be spraying water at you; they'll be spraying foam; whatever it is you just go through it and *get out of the aircraft*."

But Jim's landing was perfect. No flip, no fire, just a flash of what might have been "if their time had come".

Jim Collins experienced three forced landings in total, but never a crash. Generally, he was the one called to help with search and rescue operations for other downed aircraft. "In his later years, he was often called upon to rescue planes from awkward situations: nose into a lake, floats through the ice, or treetop landings," Brenda says. "Dad would come up with common sense solutions to extricate these planes."

The Department of Transport did not always agree with Jim Collins's strange solutions to his aviation-related problems. When he wanted to fly a float plane off the runway at the airport, he built a large dolly, set the float plane on it, and went lickety-split down the runway until there was enough lift under the wings to fly off. The dolly braked automatically as the aircraft lifted off. The technique started a trend with some of the pilots. On one occasion when he needed to ferry an aircraft with a gashed float to the mainland for repairs, he stuffed the float with inflated rubber tubes to plug the hole. However, Jim gained a bit too much altitude over the Strait and one of the tubes exploded with a terrifying bang.

One time his own float-equipped Taylorcraft, a light two-seat tail dragger, met with an accident that even Jim Collins couldn't resolve. He and some friends had flown to their cabin on Middle Ridge Lake. One of the hazards of landing on a lake is a deadhead, a log that floats just below the surface, almost waterlogged but not sunk. On this day, they hit one coming in, and unknowingly gashed a float. They tied the aircraft to a buoy because there was no room at the dock, and someone went out with a boat to pick up the pilot. The next morning, Jim walked out to the dock just in time to see the tip of the wing slip below the water. The float had filled up, and with the added weight, leaned the plane over and sunk it. Years later, Jeff went swimming there with a snorkel and mask, and found one of the instruments, but that was all – the aircraft had apparently broken up, and there were no more "souvenirs" to be found.

Big Busts and Mini Booms

Jim Collins did everything he could to keep Air Transit in the air, but still the business suffered at times from financial ups and downs.

"On the one hand, Mom would be stretching stews and everything," Brenda says, "while on the other hand it wasn't unusual for us to come down to the kitchen and see fresh lobster or crab in the sink. Dad might have made a run out to Ramea or Burgeo, and he would pick up whatever was being processed at the plants. So while other people might think of lobster and crab as a delicacy, we didn't like it much. Steak would have been a real treat. But not moose steak!"

Jim might also be hired to fly sportsmen in and out of their favourite hunting or fishing spots. He and his friends often managed to kill and share a moose during hunting season. And so the family of the entrepreneurial Newfoundland bush pilot ate well even though money was scarce.

But one winter the company found itself overextended. Eileen was seriously contemplating selling the dining table and chairs for groceries when the ice came in to the "Tickle" – the stretch of water between Portugal Cove, along the shore of the main island of Newfoundland, and Bell Island – and closed the ferry service. Bell Island, the largest island in Conception Bay, off Newfoundland's Avalon Peninsula, was at one time the largest producer of iron ore in Canada. That winter, Bell Island residents paid Jim $5 each way to get to their jobs and other commitments in St. John's. "Most times the ice came in May but that February, for the first and only time, the ice came in – and saved the dining set," Jeff adds.

The eventual closing of the mines on Bell Island was unfortunate for the people of Newfoundland but this did provide extra work in 1966 to help keep Air Transit in business. As Jim's son recalls, "A lot of these things sort of fell out of the sky when they were needed."

The Foggy, Foggy Banks of Newfoundland ...

For good reason, Jim used to refer to Ramea as the "black hole" of bad weather. Ramea is one of a group of granite-based islands located six kilometres off the southwest coast of Newfoundland between Channel-Port aux Basques and the Burin Peninsula. He could be stuck there for a week if he got caught on the ground and the fog rolled in, because it would stay. The fishermen could still go out, the coastal steamer could still come back and forth, but not the aircraft.

On one occasion, following a 10-day delay in Ramea, Jim hugged the coast to return to St. John's. He got home just in time, for the next day he was admitted to hospital for emergency surgery.

Another time, Jim was flying on floats coming back from St. Anthony, up in the north end of the island, when he found himself being chased by weather. He radioed in, "I'm on my way. I've got bad weather right behind me, but I'm just going to boot along and get into St. John's tonight." His wife then informed him she'd just got a call: there was a sick man in a small community who urgently needed to come into St. John's for hospitalization.

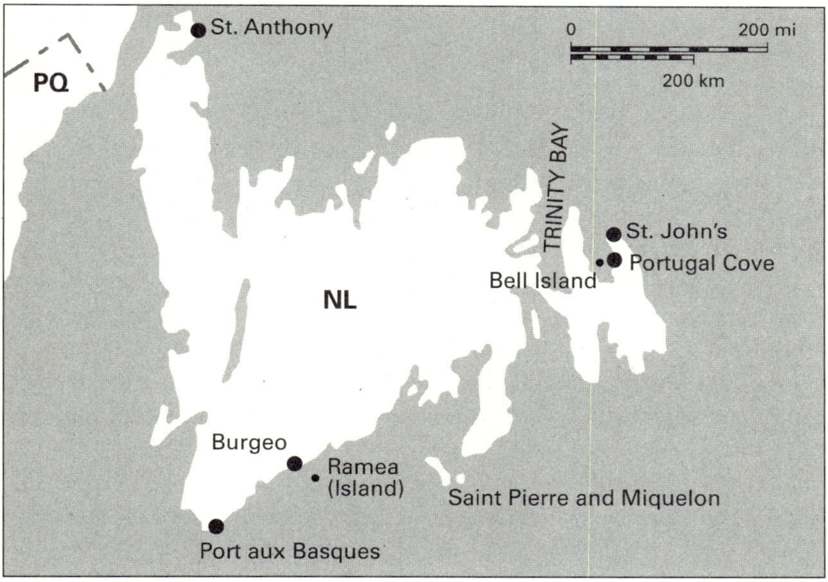

Jim Collins's Newfoundland

Jim's reply was a dejected, "Roger." Given the weather, he was very reluctant to stop, but felt obligated. The people of the outports were just getting used to this new technology of flight to bring them urgent medical care. Now, as he viewed the darkening sky, Jim hoped the nurse and patient would be waiting at the dock so it would be a quick and easy pickup.

He descended to make an ocean landing off the coast of Harbour Deep. Waves and wind buffeted the aircraft as he fought to keep it steady. Floatplanes could be easily overturned in adverse winds and it took specialized manoeuvring to remain upright. Once in past the guard rocks, he encountered the calmer waters of the harbour. He taxied up to the wharf, and there was the nurse, waiting on the dock – alone.

Jim was out onto the float before the aircraft quite came to a halt. He jumped onto the wharf and tied up. "Where's your patient? We've got weather right behind us. We've got to put him on the aircraft right now!"

"He absolutely refuses to fly," the nurse said apologetically.

Jim took a deep breath, and strode up the pathway to the patient's home. There he found a man in his 80s, covered up in bed. "Come on, Skipper," Jim urged, "We've got to get you on the aircraft and take you into St. John's. They'll make you feel a lot better there in the hospital, and then we'll bring you back home."

The ancient mariner lay back on his pillow looking frail, but determined. "No, m'son, there's no way in the world I'm getting on that h'airplane."

He'd been a fisherman all his life, the nurse explained, and a machine as strange and intimidating as an airplane was not something he was willing to risk his life on this late in the game.

"You have to," Jim said, anxiously glancing out the window. "This is the only way to get you into St. John's, and it's not going to be available to you very shortly because we've got fog moving in and you know what the fog's like here. You can operate as a fisherman in the fog, but I can't. I've got to get the aircraft off the ground."

"Yes, m'son, that's the problem," the Skipper wheezed from his pillow. "It's the 'off the ground' part that I don't like. I'm not getting in no h'airplane."

"Look, Skipper!" Jim said sharply, "the fact is, it doesn't matter. If you're out on the water fishing, or in your bed, or up in my aircraft, when your time has come, *your time has come!*"

The fisherman blinked his old eyes. "Yes, m'son," he replied. "But, what happens if we're up in that h'airplane, and *your* time comes – and mine don't?"

It was the first time Jim Collins ever admitted he'd been completely lost for an argument – but between the nurse and himself they did manage to get the patient onboard and safely to the hospital.

In those days – the 1950s and 1960s –Newfoundlanders in the outports also relied on the radio messages to bring news about family members who were in the hospital. Jeff Collins adopts his never-forgotten Newfoundland accent to cite an example. "Just a little note to Selma in Trinity, your husband John has had his operation, *huge gallstones,* they were *some big* I'll tell you, but he's fine now and he'll be home in two weeks with his jar of gallstones."

So much for personal privacy!

Change

Air Transit Limited had, over the years, owned a number of aircraft, mostly Cessnas. "Dad was a big fan of the Cessnas," Jeff recalls. "Most of his aircraft were Cessnas, except for the Taylorcraft.

When Doug Moore, a former pilot for Eastern Provincial who was a close friend of Jim Collins, came to work for Air Transit, Jim decided to buy a Grumman Widgeon, a twin-engine amphibious aircraft. It was a great ocean plane because the one big float and the two outriggers could take rougher water than a top-heavy float plane like a Cessna.

In the fall of 1968, Doug Moore, while flying the Widgeon to transport two engineers, must have hit bad weather or something unusual. The Grumman Widgeon was said to be an unforgiving aircraft, and it disappeared that day with all hands on board.

Brenda clearly recalls those bad days. "You can imagine the tension in our household through that search, it was awful. Nothing was found. Then, they were just kind of getting over the loss when some months later a fisherman brought up a life jacket and the whole search started again. Nothing more was ever found, but Dad placed a monument with the names of the three lost souls on the top of a cliff near Hopeall, Trinity Bay, where the life jacket was found.

"Dad lost an aircraft, along with a good friend and two passengers. That accident put our dad off flying, and soon after that he got out of the business."

"That accident broke his heart," Jeff states. "He never felt the same about flying after that."

Jim Collins could no longer believe the adage, "When your time has come, your time has come." He'd lost too many loved ones, out of place, and out of time.

Airport Industry

In 1968 Jim Collins turned the direction of the company from flying to servicing, and bought Hangar No. 4 at Torbay Airport in St. John's, at that time the largest hangar in Canada east of Montreal. He was back in the service station business, only now he serviced aircraft. The hangar and equipment were eventually leased to Innotech Aviation and Jim became a director of that company. As the oil business developed in Newfoundland, Jim brought in a colleague from the construction industry to design and build the Mobil Building for passengers travelling offshore to the oilfield.

When Hudson General took over from Innotech, Jim sold his facilities and equipment to Tom Collingwood. He felt good that they were in the hands of someone who, like him, would put the interests of Newfoundland

and Newfoundlanders first. It was good to see his province develop, and to realize that he had played a unique role in its aviation history.

But it had not come easy. There had been constant battles between Jim and the federal Department of Transport, which at times found Newfoundland's unique problems and Jim's weird and wonderful solutions difficult to understand. But rather than these fights and flights wearing him out, the passion kept Jim alive and thriving until he retired.

Memories

Jim Collins's time came at age 74 on December 30, 1998. His obituary in the St. John's *Evening Telegram* on January 30, 1999 summed up his efforts as both a brave Newfoundland bush pilot and unabashed businessman:

> *Jim battled weather, bureaucracy, antipathy, and many unique challenges on behalf of his airport. May he rest in peace.*

None of the Collins children went on to get their pilot's licences.

"That was one of the things, I think, that Dad was quite heartbroken about," Jeff says, "that none of us picked up that passion for flying that he had, even though we were around it all the time. But it was Dad's thing, not ours, so there was really no one to hand the company over to. We all went in completely different directions."

Paul is the only one of the Collins children who stayed in Newfoundland. Dave, the youngest, lives in Baynes Lake, B.C. His twin sister Daphne tragically lost her life in a motorcycle accident in July 1986. Jeff and Brenda are in Calgary, and Jane is in Forest, Ontario. They return as often as possible to visit their mother, Eileen, who turned 84 in August 2009.

It wasn't until Jeff left Newfoundland and started taking commercial air transport flights that he realized how much safer flying had become, compared to his father's flights in Cessnas over the foggy banks, wind-swept ocean bays, and rocky landscapes of Newfoundland. "When I was attending school in Toronto, I flew back and forth from Toronto to St. John's four times a year. It got to be just like taking the bus. And I realized how much hairier the flying was that we had all done as kids."

Brenda, too, came to understand how growing up in an aviation family affected her attitude toward flying. "I never really appreciated how much it was a part of my life until years later," she says. "I was living here in Calgary and I would go down to Toronto quite often on business. My sister Jane and her four kids lived just outside Sarnia, so I would do whatever business I had in Toronto and then take a puddle-jumper, with six or eight seats, over to Sarnia.

"I was always jet-lagged, chronically, from the work that I was doing. On one flight we took off from Toronto, and flew into a thunder and lightning storm. I was dog-tired and snoozing. We hit a pocket of air, the plane fell, and I went up in the seat and down. It woke me, and I looked around. Two women at the back of the plane were freaking out, death white, clutching each other; a man was just hanging on to anything he could grab; somebody else had his head down on his knees. I looked forward to the co-pilot and asked, 'Has the captain called for emergency positions yet?' No. 'Oh, okay,' I said, and back to sleep I went.

"Jane met me at the airport, with Katherine, her eldest daughter, who was about four years old. She said they'd seen the plane break out of a huge black cloud with lightning strikes all around. Jane thought that the devil was chasing us, and she feared what might happen to the plane, with Auntie Brenda onboard, in plain view of her small daughter. But we landed safely and the co-pilot began apologizing to everyone as they disembarked. Then he came to me and said, 'So, did you have a good nap?' I said, 'As a matter of fact, yes, thank you very much.' He said, 'You spend a lot of time on airplanes, don't you? You were the only one who wasn't upset.' I replied, 'Dad was a bush pilot.' He smiled, and nodded, 'That explains it'."

"So, half the family has this totally fatalistic view, but Jane and Paul, you'd have to sedate them to get them on the plane," Brenda says. "They hate flying.

"But, like Dad said to the old fisherman, 'When your time has come, your time has come!' And until then, all you can do is sit back, close your eyes … and enjoy the ride."

The Cooper Vane and How it Came to Be

In January, 2008, FBI special agent Larry Carr, based in Seattle, vowed to crack an intriguing mystery involving an aerial hijacking of a Boeing 727 that had taken place back in 1971. The incident took no lives, except perhaps that of the perpetrator, but the undertaking was so brave and bizarre that it continues to be a fascinating puzzle – and gained notoriety as the world's most famous aerial hijacking.

Although skyjackings are not uncommon, most are political in nature. Hijackers might demand a flight-path change to South America or the Middle East, or, even more sinister, use the aircraft as a bomb, to senselessly take many lives including those of the executors.

The 1971 hijacking, however, was different.

The Appearance of Dan (D.B.) Cooper

On the eve of American Thanksgiving, Saturday, November 27, a smartly dressed man approached the Northwest Orient Airlines ticket counter at Portland International Airport in Portland, Oregon. After stating his name as Dan Cooper, the man paid $20 cash for a one-way ticket to the neighbouring state of Washington. Flight 305 was departing at 4:35 pm, and was scheduled to arrive 45 minutes later (gate-to-gate) at Seattle-Tacoma (Sea-Tac) International Airport.

Dan Cooper boarded the Boeing 727 at Gate 52, and slowly walked down the aisle past the other 36 passengers. As the flight was only a quarter full there was no problem in getting a row of seats to himself. He chose Seat C in Row 18, near the back.

Shortly before takeoff Cooper handed a note to flight attendant Flo Schaffner. Was this a personal message? Such incidents had happened before with the comely flight attendant. She smiled politely and stuffed the note into her pocket while continuing to ensure that passengers had their seats in upright positions and belts fastened.

After the aircraft left the ground and the seat belt signs were turned off, Ms. Schaffner read the note. It was not a request for her phone number. This pleasant-looking passenger was demanding, on their arrival at Sea-Tac, the delivery of four parachutes, $200,000 in unmarked bills, and "no funny stuff." The note added the ominous postscript that he had a bomb in his briefcase.

She looked back toward Cooper to find him staring intently at her. This was serious! She opened the cockpit door and presented the note to the captain and first officer. They immediately contacted airline and airport officials, who called the FBI. The flight attendant was then told to go back and discuss the situation with Mr. Cooper.

As she approached him, he calmly looked up at her and held open the contents of his briefcase. A proliferation of red wires and sticks that resembled explosives was nestled in his bag. (At that time passengers' carry-on bags normally were not searched.) Some reports state that the captain also went back to talk to the hijacker and was shown the devices.

The Boeing flew on toward Seattle, with several more communications being passed from Cooper to the flight attendant.

Neither the FBI nor Northwest Airlines' officials could figure it out. Four parachutes! Why four? Did Cooper plan to force crewmembers or other passengers to jump with him, as hostages? Did it mean there were three other passengers on board who might be accomplices? Or was he guarding against the possibility of a defective parachute? Better give him good ones; he might take a crewmember or innocent passenger with him. Don't hand over a bundle of rags.

The contacts at Sea-Tac were given the order, but in their haste to assemble the required parachutes they included one marked with an "X", showing it to be inoperable, for training use only, that was actually sewn shut.*

At 5:24 pm, the FBI informed Captain William Scott and First Officer William Rataczak that the parachutes and money were ready. They didn't

---

* Other reports, such as *D.B. Cooper: What Really Happened* by Max Gunther, state that the parachutes specified by Cooper were two chest-type auxiliary "reserve" parachutes acquired from a sport parachute centre in Issaquah, Washington, and two standard emergency back-types from a parachuting instructor/FAA Master parachute rigger at the Issaquah drop zone.

mention that a high-speed "Recordak" copy machine had been used to record images on microfilm of all 10,000 of the $20 bills, from which they could later create a list of the serial numbers.

The Saga at Sea-Tac

The Boeing 727 landed at Sea-Tac at 5:40 pm, 20 minutes late due to circling manoeuvres, with no one aboard but the crew aware that the flight was being hijacked. Cooper sat back, allowing people to disembark until only the crew and one flight attendant, Tina Mucklow Larson, remained on board. (The pilot later recalled that the crew could have escaped from the aircraft, as they were all out of Cooper's line of sight for a time. However, he was unable to get the attention of the flight attendant in the first class section without the risk of Cooper hearing him and so the crew elected to stay aboard and see this out.)

The aircraft was quickly refuelled and the four parachutes and the bag of $20 bills, weighing 21 pounds, duly delivered. Cooper then issued the next order. Captain Scott and First Officer Rataczak were to fly him to Mexico "with landing gear down and flaps set to 15-degrees." They were to fly low and slow, no higher than 10,000 feet, and at speeds not exceeding 170 miles per hour.

The captain explained that even with refuelling at Sea-Tac the Boeing could not fly nonstop to Mexico from Seattle with the flaps down, which would make the aircraft less aerodynamic and not as fuel-efficient. Cooper, seeming to know the captain spoke the truth, finally agreed to a refuelling stop in Reno, Nevada en route to the U.S./Mexico border.

The captain then called the control tower to ask for clearance southeast to Reno with a final destination in Mexico, and the flight became airborne at 7:44 pm.

Minutes after takeoff, Cooper issued his next demand to the single remaining flight attendant. She was to open a bulkhead door at the back of the Boeing. When this door was opened, an aft staircase would unfold for passengers to exit from the tail of the aircraft. She was then told to go up front, close the first class curtains behind her, and enter the cockpit, where she was to stay with the pilot and co-pilot for the remainder of the flight.

Tina Larson complied, but before closing the curtain she quickly glanced back to witness Cooper tying something around his waist that resembled a canvas bag. It was the last view anyone had of the hijacker.

At that point the flight engineer noticed a red warning light indicating that a back door had come ajar. At 10,000 feet this was a concern but not a disaster. It didn't take much detection to figure out which door was open.

"Is everything OK back there?" the captain asked over the intercom. "Is there anything we can do for you?"

"No!" shouted Cooper.

At 8:10 pm, 24 minutes after takeoff, as the aircraft crossed the Lewis River in southeast Washington 25 miles north of Portland, Oregon, the flight crew felt a slight bump. The aircraft was trimmed to bring it back to flight altitude. Another bump. Another trim.

Might the bumps be caused by the opened stairwell bouncing in the airflow?

Whatever the case, the hijacker had jumped – to become a "folk hero" legend.

Air Traffic Control - or Not

Clifford Ammerman was employed by the FAA [Federal Aviation Administration] as an air traffic controller at Seattle Center, and was on duty Saturday, November 27, the night of the Cooper hijacking. But his recollection of the incident goes back to the night of November 23, 1971.

"A man called Eugene tower about 10:00 pm that night and asked if they were aware of an aircraft dropping flares south of Eugene in the area of Cottage Grove, Oregon. Eugene knew nothing of this," Ammerman says.

He explains that when aircraft were 30 to 40 miles away from Eugene and inbound to land, Seattle Center would normally terminate radar coverage of the aircraft and transfer communications to the manual approach co-ordinate with Eugene via hot lines. On this occasion, however, the Eugene controller called Seattle Center on the hot line to determine if they knew about the flares, or could see anything on the radar.

Ammerman was working that sector at the time, and assured the Eugene controller that he knew nothing about the flares. He checked with his supervisor for information, such as a NOTAM ("Notice to Airmen", a notification from the U.S. Federal Aviation Agency of a temporary condition, happening or hazard) that might have been issued to explain this sighting. There was nothing, and only in retrospect did this report gain importance. "In my 30 years experience as air traffic controller, nothing ever happened like that before or since – a report of someone dropping flares from an aircraft," he stated in a telephone interview with the author (and a follow-up letter dated February 6, 2004).

On November 27, Ammerman was one of the Seattle Center controllers on duty to provide service to aircraft transiting its airspace.*

---

* Clifford Ammerman explains that the Seattle Center is divided into different geographical areas (called sectors) designed for the best handling of traffic flows. One controller is responsible for each sector, and as the aircraft move from one sector to another the control, or responsibility, changes from one controller to the next. The controllers talk directly to the pilots and transfer communication to the next controller at the appropriate time. Supervisory staff would co-ordinate with the military people if a fighter escort were required, and the controller at the sector would become involved only when the aircraft were about to enter his sector airspace.

"Along came D.B. Cooper," Ammerman says, and the action began.

After speaking directly to the Boeing 727 pilot, Ammerman contacted McChord Air Force Base, which launched fighters to follow the hijacked aircraft. Two F-106s – heavy, large, delta-winged interceptors that were barely able to stay in the sky at 170 mph – took off in pursuit. The speedy F-106s had to fly in an "S" type pattern to keep from overtaking the slow-flying Boeing.

Ammerman, at Seattle Center, was now one of those charged with keeping the F-106 fighters five miles in trail as they flew 1,000 feet above the Boeing. Radar coverage was effective only in a 20 degree "cone", or 10 degrees either side of the nose; if the coverage exceeded this angle during the fighters' "S" manoeuvres the radar lock on the Boeing 727 would be lost.

Just north of Portland, an Oregon Air National Guard T-33, which could more easily trail the Boeing, joined the chase. Ammerman turned the F-106s to the east of the projected track of the hijacked airliner and then south to parallel the route. "The F-106s were climbed to 20,000 feet to save fuel and were still conveniently close in case they were needed," Ammerman recalls. "At the same time, I positioned the T-33 to 1,000 feet above and five miles in-trail of the Boeing 727 and at a speed that matched the airliner.

"With the F-106s flying higher and at a greater speed than the Boeing, an occasional orbit had to be flown by the F-106s to remain abeam of the 727."

The U.S. military operated on an Ultra High Frequency [UHF] system, continued from the times of the Second World War for purposes of secrecy, while civilian aircraft operated on Very High Frequency [VHF]. It was hoped that this dual system would prevent the hijacker from hearing what was being said to the fighters. When the Boeing reached Portland, instructions were transmitted simultaneously on VHF and UHF, forcing Ammerman to stand up and reach over to shut off the VHF before talking to the military on UHF, continually turning the transmitter boxes on and off. He kept the fighters on his frequency as long as possible.

The T-33 trailed the Boeing to Eugene, 100 nautical miles south of Portland, but at Eugene the T-33 reported low on fuel. The F-106 pilots then had to return to McChord AFB because they too were running out of fuel. This resulted in nobody following the Boeing south from Eugene, as all civilian aircraft were kept clear. An FBI agent had attempted to follow in a helicopter, but was unable to catch up to the Boeing.

Ammerman, on radar control, followed the Boeing from Seattle to south of Medford, Oregon, navigating on Vector 23 (the name of the airway that aircraft followed on this particular route from point to point), but lost

---

Further, the Seattle Center controls all of the Instrument Flight Rules [IFR] traffic that enters its airspace, whether civilian or military. The Center also provides traffic advisory service to aircraft under Visual Flight Rules [VFR] for any pilots who might desire the added safety during their flight.

radar control when the Boeing entered another air traffic control centre. At Red Bluff, California, a C-130 (a four-engine turboprop cargo plane, with short take off and landing capabilities) picked up the trail and followed the Boeing to Reno, Nevada.

When the Boeing landed at Reno, the hijacker was gone.

Because of the gap in tracking between Eugene and Red Bluff, they had lost their man. Nor had the parachutes been tracked on radar.

The day after the hijacking, Clifford Ammerman called the controller at Eugene to ask if anyone there had reported the flares near Cottage Grove, Oregon, in case the FBI might think it important. "Yes, a report was made," stated the Eugene controller. Ammerman also ensured that his supervisor knew about the flares. He was advised that the FBI would indeed be told.

Perhaps the flares had been attached to a free-falling parachute, or tied to a man's leg as he did a practice jump made far distant from the planned area of the crime, Ammerman thought. Although it was possible for parachute jumps to be made after dark, the person would have to be familiar with his landing territory. There'd be too much danger jumping into a forested area, even with an altimeter. Parachutists have used altimeters to gauge free-fall, but this instrument would be no help for landing if the height of the ground were unknown, especially in a region with trees or rocks. It would be almost imperative to have someone on the ground to guide or assist.

"The FBI said there was only a 20 percent chance of survival," Ammerman says, "but that's silly. Of course he could parachute out at night ... if he had ground help arranged. Even the way he was dressed showed he'd be confident of rescue."

With the hijacker's obvious knowledge of aviation, he would certainly have planned his escape as thoroughly as the hold-up itself.

"As well planned as the 'DBC' hijacking was," says Ammerman, "it would be odd if he didn't have something planned for his getaway, someone on the ground co-ordinating the jump, with both he and his ground-person using hand-held receivers. This would aid them in distance measuring, for example. This was over unpopulated territory, and it would be imperative to have someone on the ground."

The questions are whether the radio equipment of the time would have been sophisticated enough to facilitate such a rendezvous; also, there might have been a variance in the flight path, which would make such a plan unfeasible. Thirdly, it is not likely there would have been adequate distance-measuring equipment in portable form (while a modern-day GPS might work, such systems were not available in 1971).

Nothing came of Ammerman's report. He was never interviewed by the FBI regarding this report or his "eyewitness" role the night of the hijacking.

The Investigation

It is certain that Cooper possessed considerable knowledge of aviation, and chose the aircraft wisely. The Boeing 727 was one of very few commercial aircraft equipped with a back exit door and staircase.

It was only after the Boeing had landed in Reno that the flight crew knew for certain that he had jumped. Two parachutes had been left behind, one missing some of its nylon cords. The parachute marked "X" was also gone.

The FBI carefully bagged evidence thought to have been left by Cooper on the Boeing. This included eight Raleigh cigarette stubs, cups that had contained bourbon that might have been touched by his fingers or lips, a flight magazine found in his seat pocket, and a tie and pearl tie tack (or clip) that the flight attendant had noticed he was wearing.

So, what is actually "known" about Dan, or 'D.B.', Cooper as he came to be referred to? The initial "B" was actually a fabrication brought about by a media police report that a Daniel B. Cooper of Portland was a suspect. That person was found to be innocent, but the initials "D.B." became connected to the Dan Cooper name, or pseudonym, that the hijacker had given to the ticket agent.

And so all identified facts about this mysterious "D.B. Cooper" were meticulously recorded. He was a heavy smoker of Raleigh cigarettes; he drank bourbon; and he showed a familiarity with aerodynamics. An FBI description pegged him as a Caucasian male of about 40 years of age, 5 foot 10 inches to 6 feet tall, weighing 170-180 pounds, of average-to-good physique. He had dark brown hair, but his eye colour was unknown, as he had been wearing wraparound sunglasses. He was dressed in a dark overcoat, a dark coloured business suit, narrow black tie, white shirt, brown loafer-style dress shoes, and carried a briefcase and a paper bag.

The man had apparently parachuted from 10,000 feet at a flight speed of 196 miles per hour into the blackness of a late autumn storm. Temperatures outside the airplane were recorded at -7° F with a wind chill of -70° F. He wore no helmet or goggles, and no warm clothing could be discerned underneath his business attire. The loafers on his feet would surely have been sucked off by the windstream the moment he hit the air – if indeed they were loafers. Might they have been leather jump-boots with their high laced tops hidden by pant legs? If so, their strong structure and sponge soles would have provided excellent hiking boots to get him over the rough terrain.

It was thought he could indeed have survived those temperatures in the estimated time it would have taken him to land. Not knowing the exact ground level below him, he might have made a free fall for 10 to 15 seconds before opening the parachute canopy; with this duration of free fall, and given that the terminal speed of an outstretched body is 120 miles per hour, it would have taken him about five minutes to reach the ground. If the parachute malfunctioned, or if he landed in a tree or on a rock, the consequence would be death, or at the very least, serious injury. Hypothermia or malnutrition might have been a problem if he'd been unable to find his way out of the snow-filled forest.

The FBI employed airplanes, helicopters, patrol cars, and over 300 personnel including a force of 200 soldiers from Fort Lewis, Washington to conduct air-ground searches of the south-western Washington area. Intensive searches went on for a month. Nothing was found. "We're either looking for a parachute or a hole in the ground," stated Undersheriff Tom McDowell.

Woodland Police Chief Joe May, however, gave Cooper some benefit of doubt: "Either he's hung up in the branches of a tree somewhere and we won't find him until next deer season," the chief stated, "or he's home watching us on television, laughing his fool head off."

On February 10, 1980, a young boy named Brian Ingram, digging in the Tena sand bar on the north bank of the Columbia River northwest of Vancouver, Washington, found a soaked and discoloured package containing $5,800 in bundled $20 bills, whose numbers matched those recorded by the FBI before the money had been handed over to Cooper nine years earlier. It was the first clue ever to the D.B. Cooper mystery. Brian Ingram was rewarded with $2,760 in cash, including 15 of the original "Cooper" bills as souvenirs.

On May 18, 1980, a few months after the discovery of the money bundle, Mount St. Helens erupted, sealing the search area with volcanic ash. Clues to the mystery were now truly hidden.

Who Was D.B. Cooper?

The skies over the United States were soon well populated with similar crimes. On the morning of the same day as Cooper jumped, Saturday, November 27, 1971, three men grabbed a tow truck to get across the runway to hijack a Trans World Airlines Boeing 727 in Albuquerque, New Mexico. Armed and agitated, they stated they represented the Republic of New Africa movement and demanded to be flown to that "country". When the flight attendant lied to say the aircraft couldn't fly all the way to Africa, and added that Cuba might welcome them, they settled for being flown to Havana. The men seemed to calm down at that point, "even knocking before entering the cockpit." They were apprehended when the aircraft landed at Jose Marti Airport in Havana, relieved of two guns and a knife, and charged with the murder of a New Mexico state trooper.

The following year, three hijackers proved that it was possible to survive the jump from a Boeing 727 when they parachuted from the back staircase. None escaped the law.

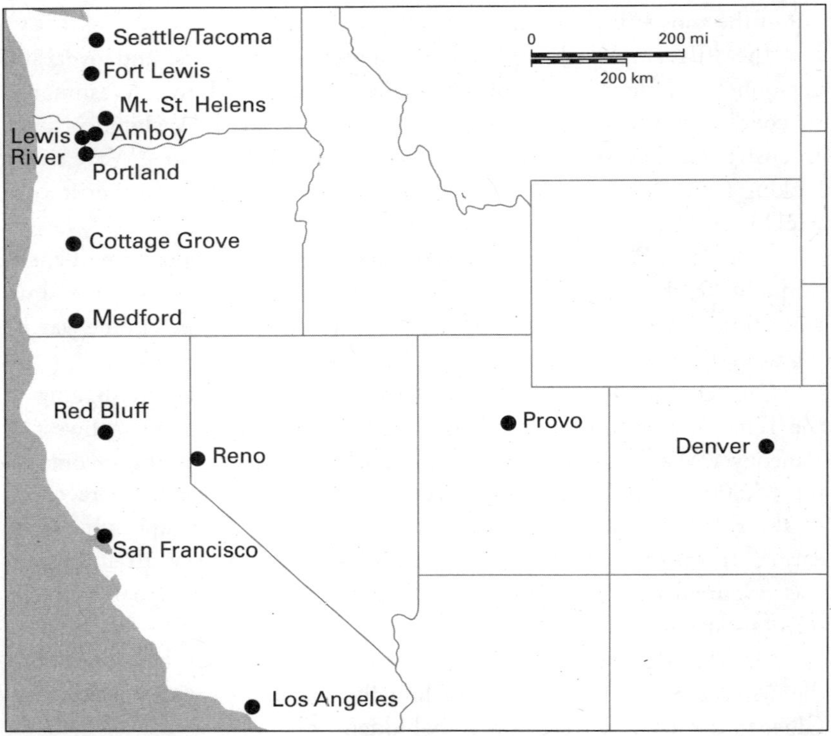

Two More Hijackings: D.B. Cooper and Richard Floyd McCoy

The most widely known incident was staged by Richard Floyd McCoy, Jr., a man often thought to be Cooper himself.* On April 7, 1972, just four months after the original Cooper incident, a United Airlines Boeing 727, Flight 855, with 85 passengers and a crew of six, stopped over in Denver, Colorado, en route from Newark, New Jersey to Los Angeles, California.

At 5:18 pm, shortly after takeoff, a passenger informed the flight attendant that a man sitting in Row 20, Seat D, was seen fiddling with a hand grenade. When the man – who, according to the passenger list, was "James Johnston" – was approached he produced a typewritten envelope titled "Hijack Instructions". A hand grenade pin and a bullet fell from the

* The possible connection between Richard Floyd McCoy, Jr., and D.B. Cooper was examined in a documentary aired on the Discovery Channel (Nov./02) titled "Flight From Justice: The Story of D.B. Cooper".

envelope, which also contained a two-page letter written on FAA flight plan forms outlining what "we" wanted. The careful wording of the letter implied that more than one hijacker was aboard the flight – it was not just a one-man show.

The man's orders were for the plane to land at San Francisco International Airport, precisely at Runway 19 Left. He named the number of people that would be allowed to approach the aircraft to deliver his request of a half-million dollars in cash, four parachutes and the return of "our" typewritten instructions.

Passengers were informed that fog at Los Angeles would require them to land in San Francisco, and the hijacker's orders were relayed to ground personnel, and subsequently to the FBI.

The hijacker's instructions were followed, including having personal baggage brought into the cabin from the cargo hold. The plane was refuelled with "six hours of fuel" as per his instructions. The man then sent out one of the passengers, a convict being escorted to prison, to retrieve the two 40-pound briefcases of money, and the parachutes. Unknown to the hijacker, the FBI special agent had attached small radio transmitters to the parachutes that would be able to track them over a six-hour period.

All passengers and some crewmembers were allowed to disembark, leaving the pilot and co-pilot and two flight attendants. Despite the request for "their" baggage to be brought up, it seemed that the hijacker was in fact operating alone, for no other person but him and the crew remained on the plane. Fuel trucks delayed the procedure as long as possible, with FBI sharpshooters trained on the aircraft hoping for a chance to knock off the hijacker. None came, and when the pilot radioed that the hijacker was becoming anxious, procedures were speeded up, and the Boeing took off after the three-hour stopover.

The hijacker ordered the two flight attendants to sit in first class seats while he took a seat near the rear of the aircraft, and used the intercom to relay orders such as to turn off the cabin lights. Strict warnings were issued that no chase planes were to be deployed; if this happened, he would set off a bomb to blow up the Boeing after he had jumped.

The hijacker then called one of the flight attendants and gave her a handwritten note outlining his demands regarding flight plan, altitude and speed: turn to the east toward Utah, climb to 16,000 feet, and fly at a speed of 200 mph. The flight attendant took this opportunity to stick the note into her purse, an action that became crucial for later identification of the hijacker.

A number of flight path orders followed, which took the Boeing along a zigzag pattern over Nevada and Utah – north, east, south. Where were they going? Near Utah, the hijacker ordered the cabin to be depressurized. He closed the peephole between the cockpit and the cabin; the second officer, however, managed to watch through a space between the floor and the cockpit door, and could see him donning a jumpsuit, helmet and parachute. By this time it was dark – the moment for the jump was approaching. The next instructions were for the captain to provide him with information on sky conditions plus wind, ground, air speed and altimeter readings. This was no layman.

Near Provo, Utah, the Boeing shuddered, indicating the back door was being opened. When would he jump? A transmitter indicated that one parachute had been deployed. Was it a decoy? Then another transmitter went off. The aircraft was flying at 14,000 feet, at dangerously slow speeds of 180 knots, just above stall speed.

In the meantime, defying the hijacker's orders, the FBI had sent out turboprop chase planes, kept at a discreet distance and flying with lights off.

After landing, the crew was instructed to leave all their belongings on the airplane while it was searched from top to bottom. All items that could have been touched by the hijacker, from seat belts to cups to cigarette butts, the in-flight magazine, food or drink containers, were taken in for testing. Especially important was the handwritten note that had been retained by the flight attendant, and which the hijacker had forgotten to ask to be returned.

Following a check with the passengers who had de-planed at San Francisco, it was learned that only the convict had seen the hijacker as he'd boarded. He'd watched the man go into the washroom and emerge in costume: wig, moustache, aviator sunglasses, and business suit. He would later become an important witness.

Police roadblocks were set up around Provo, and the muddy flats around Utah Lake were searched and re-searched. No sightings. Then a Highway Patrolman called the FBI in Salt Lake City to report that an acquaintance had once outlined to him "his great plan" for hijacking an aircraft – a plan identical to the caper pulled by "James Johnston".

The person he named was Richard Floyd McCoy Jr., who lived in a suburb of Provo. McCoy was a U.S. Army veteran who'd served two tours in Vietnam and had been awarded the DFC [Distinguished Flying Cross] and Purple Heart. He was 5 feet 10 inches tall, 160 pounds, and at that time was 29 years of age.

The convict passenger on Flight 855 was shown a head-sheet supplied by the FBI of possible culprits, and quickly identified Richard Floyd McCoy, Jr., as the suspect hijacker.

A fingerprint in *Mainliner*, the in-flight magazine, was compared with McCoy's Army print records. It matched.

Oddly, McCoy was described as a family man of the Mormon religion, who was not supposed to drink alcohol or smoke – quite different from the apparent habits of the hijacker – and was currently majoring in police science at Brigham Young University. But McCoy was also a military helicopter pilot currently employed by the Utah Air National Guard, a past member of the Green Berets, and an experienced skydiver. A smart man … but perhaps not quite smart enough.

On April 9, two days after the hijacking, the FBI moved in to arrest McCoy, catching him as he was leaving his house for a routine National Guard drill. A search warrant of McCoy's home produced incriminating evidence: a skydiving suit and equipment, a typewriter whose keystrokes matched those of the typewritten notes, a "dud" hand grenade, a .45 calibre pistol – and $499,970 in U.S. currency, just $30 short of the half-million dollars contained in the briefcases. Interesting to note, they also found a collection of clippings on the D.B. Cooper case.

The U.S. Army provided samples of McCoy's handwriting, and the suspect was asked to write out the contents of the hijack note. The handwriting was identical.

McCoy was charged with air piracy and interference with flight crew, and was indicted on April 14. A guilty verdict gained him a harsh 45-year sentence in Lewisberg, Pennsylvania's maximum-security prison, to demonstrate the court's intolerance of air piracy.

McCoy vowed to escape and did, after serving just four months. He and another inmate remained at large until August 10, 1974, when a tip enabled the FBI to trace him to a house in Virginia Beach, Virginia. When agents arrived to serve an arrest warrant, McCoy answered them with a pistol shot. Agent Nicholas O'Hara instantly returned fire with a shotgun, and McCoy's strange life came to an end.

Was McCoy indeed Dan Cooper? The similarities between the two hijacking incidences is overwhelming: their aviation knowledge; instructions being written on FAA forms; the false names, disguises of brown suits, dark ties, wraparound or "aviator" sunglasses; their method of ensuring that passengers were seated near the front of the airplane while they themselves took seats near the back, then releasing passengers at the refuelling stop when the ransom money was delivered; and the means of escape which required exceptional athletic ability and parachuting experience. A persuasive item of evidence was that the tie and pearl clip found in the Boeing that Cooper had hijacked was identical to one belonging to Richard McCoy.

In a book titled *D.B. Cooper: The Real McCoy*, co-authored by FBI agent Russell Calame, many comparisons were made to show that Cooper and McCoy could indeed have been one and the same. However, Calame told reporters at the Salt Lake *Tribune* that during police interrogation McCoy was asked, but never admitted, to being Cooper, replying: "I don't want to talk to you about it." But Agent O'Hara, who fired the fatal shot at McCoy, stated "When I shot Richard McCoy, I shot D.B. Cooper at the same time."*

* The FBI, however, did not accept Calame's theory. Furthermore, Karen Burns McCoy, McCoy's widow, successfully sued Calame and his publisher for having made the allegation.

The FBI still receives leads on the Cooper case, a mystery that has stymied them so far. More than 1,000 possible suspects have been checked out, and the data on the case exceed 100 volumes.

In June of 2000, the family of Elsie Rodgers, who'd died that May, came to police with a skull. Elsie had told her family that it was the skull of Dan Cooper, and she'd discovered it near the Columbia River while on vacation in the early 1980s. The FBI used DNA evidence originally gathered from the plane (probably from the cigarette stubs) to compare with the bone matter, but any connection was impossible to verify.

Then, in August of 2000, Jo Weber from Pace, Florida reported that her late husband, a 70-year-old antique dealer named Duane Weber, had revealed to her on his deathbed in 1995 that he was Dan Cooper. When she had pressed him for details he'd shouted, "Oh, let it die with me!" Their physical descriptions were close: Duane at the time would have been 47 years of age, 6 foot 1 inch tall, and 185 pounds in weight. He had black hair, drank bourbon and chain-smoked. He had a bad knee, which he had told her was the result of jumping out of an airplane, and he had served time in McNeil Island prison near Steilacoom, Washington, just 20 miles from Sea-Tac airport. FBI forensic investigators agreed there were many similarities, and agent Ralph Himmelsbach opened a file based on Jo Weber's statements and discoveries; however, the FBI dropped the investigation in 1998.

The FBI at that point would undoubtedly have been pleased to close the D.B. Cooper files, which had grown in lore as the only unsolved domestic skyjacking in U.S.A. history.

The Cooper Vane

The end result of the search for D.B. Cooper was the invention of an anti-hijack mechanism that came to be called the Cooper Vane. This was a spring-loaded wedge device situated on the lower rear fuselage of Boeing 727s. On takeoff, downward airflow pushes a metal plate or vane rearward, securing the ventral stairs in their upright position in such a way they cannot be lowered. When the aircraft stops the lock releases. The FAA ordered the Cooper Vane to be installed on all Boeing 727s.*

* The actual Boeing 727-051 used in the D.B. Cooper hijacking was delivered to Northwest Orient Airliners on April 22, 1965 and registered as N467US. It was sold to Piedmont Air circa 1982 and re-registered as N383N, sold to Key Airlines in May of 1985 as N29KA, and then acquired by WorldCorp after Key went under in 1993. According to a former Key employee, "The Cooper story was well known to us, as was the fact that we owned the 'Cooper' B727." The aircraft's last flight came in 1993, when it was delivered to the scrap yard "to be turned into beer cans."

Further regulations were put in place in 1972 that required all air luggage to be run through security checks.

A New Twist on an Old Mystery?

A Canadian link to the Cooper mystery was recently acknowledged by FBI Agent Larry Carr, the Seattle-based agent assigned to the case when the FBI formally reopened the investigation in December 2007. As Carr doggedly sought out new or formerly overlooked information he came upon a Canadian comic book published in French that depicted a hero named Dan Cooper, a Royal Canadian Air Force pilot. Published in the 1960s, the comic book's handsome hero successfully battles earthly and alien enemies. One cover illustration shows him preparing to make a daring parachute jump.

Meanwhile, a Calgary, Alberta-based pilot (who insists on anonymity) presents a quite different perspective regarding the Cooper case. "Find out what small aircraft flights took off from Sea-Tac airport within 90 minutes of when the 727 departed, and you'll have your man," he states.

After the passengers had been allowed to disembark at Sea-Tac, and the money and parachutes had been brought on board as directed, only the basic crew remained on board with the hijacker. The Calgary pilot's theory is that D.B. Cooper never left the airport.

All indications are that Cooper had thorough knowledge of airline operations, ground operations, and routing, and specifically of the mechanics and aerodynamics of the Boeing 727. He knew that this was one aircraft that could actually fly with the tail door open a bit. If it is possible that the door to the stairs were opened *before* takeoff (three feet or so, just a crack), it brings an altogether new perspective to the mystery.

"It would not be difficult to short-circuit the warning lights by using a short cable with alligator clips, or even a piece of paper, to interfere with the contact points," the Calgary pilot surmises. "Cooper might have done this himself, knowing that with his orders to the captain to fly no higher than 10,000 feet, the aircraft would not have to be pressurized."

So, if the door had been opened about three feet *before* takeoff, either by him or the flight attendant, he could sit, or lie, on the end of it, facing forward with his back and parachute outward, then pull the handle and open the chute.

"He would do this at 200 to 400 feet above ground level, with the 727 in a rapid climb," says the pilot. "The canopy would fall into the slipstream, open up, and pull him out of the aircraft – free, clear, and with no part of the aircraft behind him to tangle the chute. This is how freight drops are made from aircraft like the C-130.

"His body, as he exited, would act as a pendulum to pull him down below the canopy of the parachute. His landing would be on the flat, solid ground of the airport – not on the spires of trees or snow-covered rocks as supposed."

Now for the escape. The man was a pilot himself, as shown by his knowledge of the flight characteristics of the 727, and perhaps had a small aircraft such as a Cessna 172 parked at the airport ready to go. "He lands on the grass at the end of the runway, packs up the parachute, and makes his way to the parking area – staying in the dark well away from the actual runway – to where his aircraft is parked. He gets into the aircraft, calls Air Traffic Control, and files a flight plan to somewhere. In the dark no one could verify the registration he'd give."

Then, he's gone – into legend and history.

"The question becomes, of course, whether the door was opened before or after takeoff; the difference is when he last communicated with anybody. If it is certain that the bulkhead door was opened 15 minutes *after* takeoff by the flight attendant, this means he didn't bail out immediately at the airport. But, the possibilities are there."

Admittedly this theory seems to run counter to other parts of the story – for instance, the captain speaking with Cooper over the intercom (reported to have occurred some time into the flight) and the bump (believed to be the stair being lowered, again sometime into the flight). Neither of these occurrences would have happened at only 400 feet over Sea-Tac. But perhaps timelines were muddled with the excitement of the situation. Certainly the theory does offer both credibility and intrigue.

Now, the money found on the riverbank? "A decoy," states the Calgary pilot, "to draw attention away from who and where Cooper was."

What about the Raleigh cigarette stubs, his penchant for bourbon, the tell-tale tie and pearl tie clip left behind? "Red herrings," surmises the Calgary pilot.

So, if he did escape with the money, why were the numbers on the currency never traced? "Money disappears all the time into the underworld," is the reply. "You take it to an offshore bank, deposit the cash. There it is." Were the serial numbers on the bills really copied? "It was a pretty quick job if it was, considering the kind of copying apparatus that would have been in use at that time. And," adds the Calgary pilot "the investigators might be trying to cover their own butts by saying it had all been copied."

Was there an accomplice on the ground to co-ordinate his jump? "Not likely. How do you co-ordinate something like that at night? And Cooper wouldn't know the exact routing the aircraft would fly." The Calgary pilot then adds with a smile, and a reminiscence of his own military background: "Why practise something you have to do right the first time?"

TOP LEFT: Frank McLeod (*from Deadmen's Valley*)

TOP RIGHT: Willie McLeod

ABOVE: The remains of Albert Johnson's cabin at Rat River after it was blown up by dynamite. To the astonishment of the pursuing posse, Johnson survived the blast.

LEFT: Wop May (left), pilot; and mechanic Jack Bowen with the Bellanca CH-300 Pacemaker (CF-AKI) used in pursuit of Albert Johnson.

TOP: The body of Albert Johnson lies on the frozen Eagle River, shot by members of the posse who fired from both sides of the riverbank.

ABOVE: The wreckage of CF-FYN, the Cessna 180 flown by Blake MacKenzie. Even though the crash occurred on January 5, the wrecked aircraft was not found until August 14, 1962 (*from* "Deadmen's Valley – Pictures of the Nahanni).

TOP: The only memorial of CF-CPC is the commissioned artwork "Toward Their Final Destination" by B.C. artist Dan Ryan; the painting can be seen at the South Terminal of Vancouver International Airport (*from* "The Unfair Weather of Mount Fairweather").

LEFT: George Maude and Shirlee Smith Matheson peruse the "Book of Broken Dreams", a massive collection of aviation stories. The book, bound in wooden covers, measures 27 inches long by 20 inches wide by 5 inches thick, and weighs 42 pounds (*from* "The Book of Broken Dreams").

TOP: The wreckage of the Fairchild 24W46 tells the sad tale of an inexperienced pilot brought down by the treacherous weather conditions that prevail on British Columbia's wild coastline (*from* "The Book of Broken Dreams – The Loggers' Discovery).

ABOVE: U-537 in Martin Bay, Labrador (*from* "Nazi Toads and Other Wartime Invasions").

TOP LEFT: *Kapitänleutnant* Peter Schrewe, commander of U-537.

TOP RIGHT: Scientific Officer Dr. Sommermeyer on board U-537.

ABOVE: Weather station WFL26 at Martin Bay, Labrador.

TOP: Klaus Sonnenberg with his Cessna 172 in Grand Manan (*from* Soaring with Seagulls).

ABOVE: Klaus and his Atlantic Charters aircraft frequently find themselves visiting out-of-the-way locations, such as Nuuk Airport in Greenland; the absence of de-icing equipment at the airport adds to the complications of flying there.

RIGHT: Klaus Sonnenberg at the controls of the Atlantic Charters' Piper Navajo Chieftain Panther.

TOP: Jim Collins, second from right, with the Norseman used by Newfoundland Aero Sales and Service for the mail run up the northern peninsula. On Collins's left is Joe McGillivray, a former air force pilot admired for his ability to "fly anything" (*from* "When Your Time Has Come").

ABOVE: The Air Transit hangar in St. John's, Newfoundland. Air Transit Limited, the aircraft charter service founded by Jim Collins, connected fish company owners with plants and suppliers along the coastline of Newfoundland and Labrador. In a province where the sea had traditionally served as the "highway", roads to rural coastal communities were virtually non-existent, and Collins and Air Transit provided a much-needed service.

RIGHT: Likenesses of D.B. Cooper circulated by the FBI. These sketches were released for the first time in early 2008 (*from* "The Cooper Vane and How it Came to Be").

BELOW: A model of the Boeing 727 showing the rear staircase used by D.B. Cooper, Richard Floyd McCoy and other hijackers as their in-flight escape exit. *"Cooper possessed considerable knowledge of aviation, and chose the aircraft wisely – the Boeing 727 was one of very few commercial aircraft [that was] equipped with a back exit door and staircase … and could actually fly with the tail door open a bit. … The back staircase is way below where you'd be affected by turbulence from over the wings."* (abridged from the text)

TOP: The wreckage of Canadian Pacific Air Lines Flight 21. *"The tail section had completely separated from the fuselage, and the remains of the aircraft and its passengers lay scattered over a 900-foot-wide area. It was a terrible sight ..."* – from the text (*from* "Explosion at Puntzi Mountain").

LEFT: *"It is totally destroyed,"* reads the cryptic note on the Department of Transport accident report regarding the Douglas DC-6B (CF-CUQ). *"Remove from register."*

RIGHT: Matt Murphy at the sod shack, Musk-Ox Lake, NT, photographed in 1941 (*from* "The Murphys of Musk-Ox Lake"). For Neill Murphy, this was home between 1937 and 1952.

BELOW: The Associated Airways Bellanca Skyrocket piloted by Max Ward following the crash at Beechey Lake. *"The Bellanca is a little bit 'blind' when you come down for a landing, and all of a sudden I saw this hill in front of us. I yelled to Max, 'Pull it up!' He did, but we were too far down at the time. We hit this hill and it kicked us up into the air … The Bellanca landed with one wingtip pointing skyward and the other touching the ground, with the disconnected wheels scattered out on the snow."* – Neill Murphy, from the text.

RIGHT: Jacqueline and Terry Murphy at the Glowworm Lake outcamp west of Musk-Ox Lake; the bouncary of Nunavut is about 10 miles to the south.

TOP: Members of the Perry River Expedition pose in front of their chartered Associated Airways Avro Anson. From left to right: Harold Hanson, U.S. Wildlife Service; Paul Queneau, National Geographic Magazine; Doug Ireland, pilot; Peter Scott, expedition leader.

ABOVE: Neill Murphy on the Gulf Oil Beaver CF-GQN in Deadmen's Valley, Nahanni.

TOP LEFT: Preparing for the long flight: a deHavilland DHC-6 Twin Otter in the Kenn Borek Air hangar in Calgary immediately prior to taking off on the first mid-winter medevac flight to the South Pole in April 2001 (*from* "Kenn Borek Air – Anywhere! Anytime!").

TOP RIGHT: Mark Cary, Norman Wong and Sean Loutitt at Amundsen-Scott South Pole Station.

ABOVE: Warming up the Twin Otter with Herman Nelson heaters at the South Pole.

RIGHT: The temperature gauge tells its own story.

TOP: Leaving Rothera, 2003.

ABOVE: Sandi and Sean Loutitt. Sean is wearing the Meritorious Service Decoration presented by the Governor General, The Right Honourable Adrienne Clarkson; as the citation noted, *"[their professionalism] brought great credit to Canada, both nationally and internationally."* A second successful midwinter medevac flight to the South Pole station was captained by Sean Loutitt in 2003.

LEFT: Back to business: Sean Loutitt, Vice President, Operations, in his office at Kenn Borek Air headquarters in Calgary.

ABOVE: The Alta Flights (Charters) Inc. Cessna 414A that was destroyed when it struck a ridge in the Highwood Range on September 23, 2003 was not the usual aircraft used in the cargo and mail run from Cranbrook to Calgary (*from* "So Others May Live").

RIGHT: Garry Wutzke, Commander of CASARA Zone 2 (Southern Alberta).

FACING PAGE, BOTTOM: Spotters found traces of the crash of Alta Flights' Cessna 414A (C-GVZE) where it had hit just 50 feet below the top of an 8,900-foot ridge. *"Most of the airplane parts went over the other side. The biggest visual item was the orange ELT sitting on the ledge of the mountain, and a burn spot on the mountainside that was being investigated when they found the ELT."* – Garry Wutzke, from the text.

ABOVE: Finding wreckage from the air can be a tremendously frustrating challenge. Can you see the crashed airplane in this CASARA photo? Even with the drawn-in circle, it's difficult to spot the downed craft – the Cherokee that crashed near Porcupine Creek in Kananaskis Creek, May 8, 2004. There was one fatality, but one person was saved.

LEFT: Hal Tetz finishing a 100-hour inspection on a Be helicipter at Eureka, Nunavu latitude 85 (*from* "The Rev. Hal Tetz – Helicopter Pilot").

BELOW: Range Airways' chartered Canso and Norseman docked at Echo Bay, Great Bear Lake, Nunavut. In the summer of 1964 Hal Tetz found himself on the right seat of the amphibious Canso, with Jimmy Lougheed as captain, to transport fishermen from Edmonton to Great Bear Lak and back. The Canso *"was a big slow old bird - you had to really fly that thing, it was hard work."*

Significance of the flares? "Only one person reported them. They could have been from anything, such as flashing strobe lights which, at that time, were just beginning to be installed on aircraft."

How about the vortex from the aircraft? Wouldn't a jumper be affected by that, and be buffeted all over the place? "The main vortex would be from the wing tips and he would be well clear of that area. Sure, he'd be blown around some, but people jump from airplanes all the time, such as C-130 Hercs, deHavilland Buffalos, or DC-3s out the side door. The back staircase of the 727 is way below where you'd be affected by turbulence from over the wings. Not a problem, actually."

The understated conclusion is, "The person we know as D.B. Cooper was an exceptionally bright person with a high degree of technical skills."

As for some evidence of identity, the Calgary pilot states, "When I asked the suspect if he was, in fact, 'Cooper' he did not deny it. He just said, 'Do you want to see the parachute burns?' By now they would have healed, long ago actually, and there would be no marks left."

The "suspect" for obvious reasons must remain anonymous, with a federal air piracy charge still current. And of course, legendary heroes and anti-heroes must maintain their mystique. All that can be promised at this time is that the identity of "D.B. Cooper" would be revealed to this author after the hijacker's death. As "Cooper" is now about 73 years of age (in late 2009, which would have made him about 37 when he performed his death-defying feat), the day of revelation for both him and his acquaintance, necessarily referred to here as simply "the Calgary pilot", might not occur for some time.

The Fans and Foes of D.B. Cooper

Several books have been written about the D.B. Cooper saga*, as well as a song or two and a musical stage play, *The D.B. Cooper Project: A Cabaret Musical of a Hijacking Myth* by John Orlock. A film, *The Pursuit of D.B. Cooper*, was produced by Universal Studios in 1981, starring Robert Duvall, Treat Williams, Kathryn Harold, and Ed Flanders.

Prominent journalists considered the D.B. Cooper legend to be among the most interesting stories they ever covered. One such person was Wick Temple, who, during his long career had worked on major news and sports events throughout the United States and had held executive positions

* *D.B. Cooper: What Really Happened?* by Max Gunther; *D.B. Cooper: Dead or Alive* by Richard Tosaw (Tosaw has actually listed all 10,000 serial numbers from the $20 bills given to Cooper, and offers a $10,000 reward to anyone who finds any of the ransom money); and *NORJAK: The Investigation of D.B. Cooper* by Ralph Himmelsbach, a retired FBI agent who attempted to chase the Boeing by helicopter and spent the rest of his working life reviewing the case and following up on leads.

ranging from bureau chief to senior editor to vice president of Associated Press; when he died, the headline on his obituary in the Seattle *Times*, published February 3, 2003, focused on the scoop of his career: "News exec covered D.B. Cooper story."

"Mr. Temple happened to be at the Seattle-Tacoma International Airport at the time," states reporter Christopher Sullivan. "He recalled later, 'I dictated material from Sea-Tac until the plane took off … . The skyjacker, as everyone knows, bailed out from the tail stairs of the 727, and we had that memorable quote from the FBI that 'he's probably screwed into the ground up to his elbows.'"

Even now, almost 40 years later, "Cooper" items are being produced that range from t-shirts (imprinted with the logo "*D.B. Cooper, Where Are You*?") to coffee mugs, posters and other memorabilia. A number of restaurants relate to D.B. Cooper, as does one in Salt Lake City that holds "Jump Night" parties with prizes of a round-trip ticket to Seattle or parachuting lessons.

Such is the mystique surrounding the man and his caper that since 1976 the Ariel Store and Tavern in the village of Ariel, Washington, near the spot where Cooper is believed to have jumped, continues to throw annual parties in his honour on the Saturday following American Thanksgiving. Frequently a visitor to Ariel will claim that he, or a relative, is D.B. Cooper, especially following public airings of the movie or documentaries relating to the Cooper incident.

Many miles of the Columbia River have been dragged for Cooper evidence, and even a small submarine was deployed to check out the bottom of nearby Lake Merwin, pinpointed by FBI computers as his likely landing place.

What factors fashion a hero from a villain? Getting away with the gold? A complete disappearance that leaves no clues? Not hurting anybody except insurance companies or large corporations? The fact that air piracy charges remain on file with the U.S. District Court against Cooper has not lessened his "hero" status.

The Mystery Renewed

In 2008, FBI Agent Carr of Seattle began his investigation into the Cooper mystery by issuing on the Internet (www.fbi.gov) a summary of all information gleaned to date, complete with artists' sketches of the man and photos of artefacts left behind in the aircraft. These items might still provide traces of DNA: the black J.C. Penney tie, which Cooper had removed before jumping; cigarette butts; a parachute and its bag left behind in the aircraft; or some of the recovered $20 bills.

It is Special Agent Carr's belief that Cooper was not a paratrooper, nor even an amateur skydiver, and in all likelihood never opened his parachute. "No experienced parachutist would have jumped in the pitch-black night, in the rain, with a 200-mile-an-hour wind in his face, wearing loafers and a trench coat." It is also difficult to imagine that he had "ground crew" assisting his caper. But, any hypotheses will be heard: the gates are open.

In March 2008 a parachute canopy was discovered by some children playing in the woods near Amboy in southwest Washington State. The canopy, found when the children's father was repairing a road, was pulled out of the ground as far as possible, and the ropes cut with scissors.

So what were the chances that this could finally be the clue leading to the solution of the mystery? This was indeed an area where Cooper could have landed and if this item was from a United States Navy NB-6 parachute, the type used by Cooper, it could provide an intriguing clue that Cooper had survived the jump. Newspaper articles headlined this potential new clue to an old mystery, and numerous explanations surfaced as to why the deployed parachute would be buried. Perhaps Cooper had realized that an abandoned chute would soon be spotted by follow-up air searches for him. The ragged parachute was examined by Earl Cossey, a parachute expert who had actually provided the chutes that the FBI issued to Cooper in 1971. Alas for the theory, the newly excavated parachute dated back to the era of 1945 and was made of silk, whereas the ones Cossey had provided to the FBI and thence to Cooper were made of nylon. Another potential clue checked off.

D.B. Cooper was the master of a daring heist climaxed by a plunge of 10,000 feet into a black forest in the dead of night. Or was he? The incident has since become a mystery that neither time nor the FBI have yet been able to explain, but which promises to bring forth even more intriguing theories. Keep tuned ...

Saboteurs

While such occurrences are relatively rare, airliners are sometimes the target of madmen who bring death and destruction to few or many by use of explosives. In an aircraft there is no place for victims, intended or otherwise, to hide from the devastating force of dynamite set off when love has turned to hate, or perceived injustices cloud common sense, and revenge overtakes the human mind.

The Guay Affair

An incident that rocked Canadian skies occurred on September 9, 1949, when a Canadian Pacific Air Lines DC-3 (CF-CUA) exploded near Forrestville, Quebec on a flight from Montreal to Baie-Comeau. A short time earlier, the aircraft had stopped over at L'Ancienne-Lorette, a suburb of Quebec City. Among others, a passenger named Rita Morel had come on board. A package was also brought on as cargo and placed in the forward baggage compartment.

Rita Morel's husband, Joseph-Albert Guay, saw her off on the flight, after purchasing a $10,000 flight insurance policy on her life, which added to a $5,000 policy in her name bought seven years previously.

The marriage of Rita and Albert, as he was usually known, had not been going well. Debts had piled up, especially after the birth of their baby, and although Albert was a smooth salesman in the jewellery and watch business, sales were down. Lately, Albert's sorrow had been lightened by a 19-year-old waitress named Marie-Ange Robitaille, whom he was seeing on the sly under the name of "Roger Angers". In fact, he had recently set her up in an apartment and offered her an engagement ring. When Rita discovered his infidelity, she'd confronted Albert and Marie-Ange. Marie-Ange promptly quit him when she discovered he was married.

Divorce was impossible in a Catholic marriage in Quebec in those days, and so Albert seethed while contemplating ways to get rid of his wife.

He considered poison but that was too easily detected. Air transport, on the other hand, was still considered a bit risky in 1949, and accidents were not uncommon. Albert flew fairly often with his job. Perhaps he could convince his wife to take a flight to pick up some merchandise for him. If she agreed, he could arrange to have a bomb placed on board, set to go off over the St. Lawrence River. By the time investigators brought up the aircraft and the bodies, evidence would be so mixed up that nothing could be traced.

Albert contacted a trusted friend, Généreux Ruest, who was a clockmaker, and asked him how such a device could be made. Ruest brought his sister, Marguerite Ruest-Pitre, into the plan. It was Marguerite who'd arranged the initial meeting between Albert and Marie-Ange; eager to help, she purchased the dynamite at a hardware store, stating it was to be used in clearing stumps from a field. Monsieur Ruest designed a bomb using the dynamite, batteries and a clock, and Marguerite duly delivered the package to the airport as cargo for the same flight that Rita was boarding.

The departure was a bit delayed, however, and so the bomb did not go off as planned over the river. Forty-one minutes after takeoff it exploded over Sault-au-Cochon near Saint-Joachim on the north shore of the St. Lawrence. All 23 persons aboard were killed: the four crewmembers and 19 passengers.

Marguerite attempted suicide 10 days after the event, and while recovering in hospital, confessed to her part in the crime. Albert was arrested three days later, having already put in a claim for his wife's insurance policies. But when charges were laid, both Généreux Ruest and Marguerite Ruest-Pitre proclaimed innocence. Marguerite said Albert had told her the package she was asked to deliver to the aircraft contained a statue, while Généreux said he believed that the bomb he'd made was for blasting tree stumps.

Joseph-Albert Guay was tried for the crime in February 1950. He then made a statement incriminating Marguerite and Généreux, saying they had knowingly assisted him in preparing the weapon for a deadly mission.

Guay was convicted and sentenced to hang. At his execution on January 12, 1951, at age 33, his last recorded words were *"Au moins, je meurs célèbre!"* – "At least I die famous."

Généreux Ruest and Marguerite Ruest-Pitre also "died famous". Généreux was arrested on June 6, 1950, tried and convicted in November, taken to the gallows in a wheelchair (he suffered from osseous tuberculosis), and hanged on July 25, 1952, at age 54. Marguerite was arrested on June 14, 1950, tried on March 6, 1951, convicted, and hanged on January 9, 1953 – the 13th and last woman to be hanged in Canada.

Copy-Cats

On November 1, 1955, a United Airlines DC-6B (N37559) exploded in midair near Longmont, Colorado.

Flight 629 began in New York, landed en route in Chicago, and then flew on to Denver where it had a crew change. Shortly after takeoff, air traffic controllers at Denver's Stapleton Airport saw two bright flashes of light in the sky to the northwest, which quickly plunged to the ground. All nearby flights were contacted and accounted for – except United Airlines Flight 629.

Ground searchers, following directions from local residents who had heard explosions and seen debris falling from the sky near the town of Longmont, came upon a grisly sight: the DC-6B had crashed, spilling its parts and contents over six square miles of countryside. All 44 persons aboard (five crewmembers and 39 passengers) were dead.

It didn't take long for investigators to zero in on the cause, with a strong odor of explosives emanating from items in the number four baggage compartment leading them to the epicentre of the carnage. Now, of all the passengers, who had just purchased flight insurance? One was Daisie Eldora King, a 53-year-old Denver businesswoman who was on her way to visit her daughter in Alaska. Surely this person couldn't have planted the bomb – but somebody might have wanted her dead to collect the insurance.

Inspection of King's bags brought forth a newspaper clipping about her son, John (called Jack) Gilbert Graham who, in 1951, had been arrested for forgery. Further investigation found that a Denver restaurant owned by Mrs. King, the Crown-A Drive In, had been damaged by an explosive. Graham, who managed the restaurant, held the insurance policy for the building and had promptly collected on it.

A search of the Denver home and vehicle belonging to John Graham, who was married with two children, revealed wire and bomb-making equipment identical to that found in the aircraft. Slowly, the sad saga unfolded.

John Graham, after offering conflicting alibis and excuses – including a tale of an unhappy childhood – finally admitted to formulating the plan to kill his mother and collect a $50,000 insurance policy she held. Further, he was the beneficiary of her will and other insurance policies so she apparently was worth much more to him dead than alive. As revealed by Graham's wife, he had personally wrapped a gift for his mother that morning. When Daisie

King prepared to check in at the airport for her flight to Alaska via Seattle, Graham handed her the gift-wrapped *bon voyage* package, which she placed in her luggage. Unknown to her, or anyone else on the flight, his parting gift to his mother was a bomb.

Daisie died in the crash, along with all fellow passengers and crew. Graham was tried and convicted on a charge of first-degree murder. It was thought that he had copied the plot of Joseph-Albert Guay. Unfortunately, it *did* replicate the Guay affair, right to the perpetrator's execution for his shameful deed.

An unremorseful John (Jack) Gilbert Graham was sent to the gas chamber at Colorado State Penitentiary on January 11, 1957. "As far as feeling remorse for these people, I don't," he said in his parting statement. "I can't help it. Everybody pays their way and takes their chances. That's just the way it goes."

Explosion at Puntzi Mountain

Puntzi Mountain, a beautiful site near the historic Gang Ranch, approximately 120 miles west of Williams Lake, B.C. on the Chilcotin highway, was once the location of a USAF Second World War military staging strip. In November 1962 the site was taken over by the RCAF and renamed 55 Aircraft Control and Warning Squadron. It later operated as a weather station providing air surveillance information to NORAD [North American Aerospace Defense Command] and serving as an emergency recovery base.

The explosion of an airliner on July 8, 1965 over the site proved to be their worst emergency recovery assignment. The Unit Ground Search & Rescue Team at Puntzi Mountain was sent to assist the RCMP at an aircraft crash site 10 miles north of Gustafsen Lake near Dog Creek in central B.C., where they were told a Canadian Pacific Air Lines Douglas DC-6B (CF-CUQ) had exploded in mid-air, at 16,000 feet. It would be B.C.'s worst air disaster since a Trans-Canada Airlines North Star had crashed into Mount Slesse in December 1956, killing all 62 crew and passengers; like its predecessor, the cause of the CF-CUQ crash became a mystery of misery.

"It is totally destroyed. Remove from register," reads the cryptic note on the Department of Transport accident report regarding CF-CUQ.

Smoke from the wreckage could be seen 25 miles distant, through the lightning and heavy rain that permeated the area that summer evening. The tail section had completely separated from the fuselage, and the remains of the aircraft and its passengers lay scattered over a 900-foot-wide area. It was a terrible sight as rescuers attempted to recover the 52 bodies and gather evidence that would hopefully prove the cause of the disaster.

Canadian Pacific Flight 21 had departed from Vancouver International Airport at 15:30 hours on July 8 for a scheduled flight to Prince George, Fort St. John, Fort Nelson, B.C.; and Watson Lake and Whitehorse, Yukon. At 16:40, Captain John Alfred Steele, an ex-RCAF pilot, had noted turbulence ahead and advised that he was altering course slightly to avoid the weather. Accompanying him in the cockpit were First Officer Warner Murray Wells and Second Officer Stanley E. Clarke. Crewmembers were flight attendants Ernest Wentzel Soural, Sue Heinrich and Marlene Brauer.

At 16:55, just 25 minutes before the flight's ETA at Prince George airport, three desperate "Mayday" calls were made. Then, nothing.

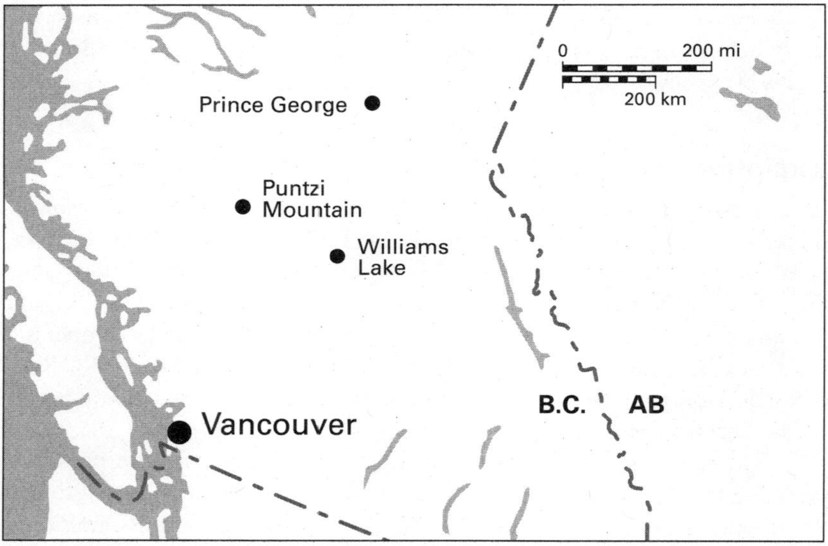

Tragedy at Puntzi Mountain

When area fire spotters noticed smoke they sent out Slim Sherk, a pilot working for the B.C. Forest Service at nearby Williams Lake, to check out the source in his Piper Cub. Sherk viewed an unforgettable sight. "The fuselage was smoke and flames. The wings were collapsing from the heat but the plane was relatively intact except for the tail section," he reported to the Vancouver *Province*. The broken-off tail section lay a half-mile distant from the main fuselage section, which had landed intact.

"There were bodies and debris all over. I counted 20 to 24 bodies until I couldn't find any more. There was no sign of life."

Sherk circled for nearly an hour as he threw out toilet paper trails to indicate locations of the bodies.

An RCAF Albatross quickly flew to the site, bringing a three-man para-rescue team who jumped to the disaster to search for victims, and hopefully find some alive. What they found were 24 bodies strewn within a 150-yard radius of the crash site, and others inside the front section of the aircraft with seat belts still attached.

Personnel from Comox RCAF air-sea rescue units, Canadian Pacific, the Department of Transport, the Forestry Service and the RCMP joined in the investigation. Vancouver Coroner Glen McDonald was assigned to the case, and acted as spokesman for the group of investigators.

It became obvious that the explosion had occurred in the tail section of CF-CUQ, which was blown over a wide area while the wings remained attached. Initially, investigators thought the crash could have resulted from one of three factors: decompression, fire, or lightning. It could also have been set off by a spark from a small gas-powered heater used for de-icing in the rear of the plane. None was proved to be correct.

The Vancouver *Province,* on July 14, 1965, offered a graphic description of the wreckage. The explosion was so violent that the stainless steel lavatory appliances in the rear washroom, on the left side of the airplane about 25 feet from the tail, had crumpled like tissue paper. The lavatory floor was blown out, as were the bulkhead wall and fuselage. Rivets were sheared from the exterior of the aircraft. Twisted pieces of the toilet and wash bowl had folded like a tulip.

All these fragments, along with all other parts, were sent to RCMP crime labs in Vancouver for analysis, and then to a Vancouver airport hangar for re-assembly of the rear section in an effort to determine the cause of the crash.

Special investigator Dr. Warren Lovell from Seattle, Washington, a pathologist who had investigated more than 150 crashes involving 2,000 deaths, was brought in to help examine the evidence. He ascertained that the "accident" had in fact been caused by a bomb.

Had an unidentified saboteur, with access to explosives and perhaps a motive to stage a suicide mission, placed a high-explosive device in the washroom? Such an incredibly cruel and selfish act had taken the lives of 46 passengers and all six crew.

The passengers had included corporate executives, families, miners, and summer students going north for jobs to pay for their next year's college tuition. All personal records were checked. Although only one had taken out an insurance policy, this was examined carefully during the inquiry. In fact, this event increased requests from the Canadian Airline Pilots Association to ban the sale of pre-flight insurance from vending machines at airports. "The easily available insurance is a temptation to would-be bombers," stated CAPA first vice-president Captain Neil White.

Ormond Hay, a miner, was one of the luckless passengers. "Ormand always took a seat at the back," recalls Jean Buck, who knew him from

Uranium City. "He felt it was safer at the back of the plane, but the irony was that the bomb was set off in the washroom right near his seat."

Certainly there were people aboard who had access to explosives from their former or current jobs, such as a powder man en route to a job in Cassiar, B.C., who, it was said, could have been carrying detonators. This was vehemently denied by his employer, who stated that explosive workers know the detonators are never shipped by air, but carried in special rail cars or by truck. Alternatively, some passengers might have been harbouring personal or work-related problems, but this accident affected so many and was so serious that quick analysis was impossible.

Although explosives usually leave a thin coating of nitrates on nearby surfaces and none was initially found, such evidence could have been washed off by the heavy rain that drenched the crash site following the explosion. Chemical tests using a microscope and chromatography were employed to seek out any traces of nitrates on under-surfaces or in small cracks in parts of the wreckage. By July 24, investigators reported that gunpowder residue had been discovered in the wreckage. "The traces of potassium nitrate and carbon found on a wall of the rear lavatory compartment of the plane indicate use of a low-velocity explosive such as gunpowder or stumping powder," reported the Vancouver *Province*.

With the discovery of the gunpowder, a public inquiry was now considered unnecessary. The case was turned over to the RCMP, who were unable to verify that the explosion was deliberately detonated, or that a person had brought a detonator such as a blasting cap on board with the intent to blow up the plane.

"Anyone with a few cents and a grade eight education could have caused the explosion," commented a local explosives expert. "Gunpowder, also called black powder, can be made by simply mixing potassium nitrate, sulphur and carbon. Anyone can buy the potassium compound – saltpetre – at any chemical store for $1 to $2 a pound. ... all you have to do is put it in a sealed lead pipe."

He added that stumping powder, somewhat more powerful than gunpowder, could be purchased by any "farmer or logging operator without any permit," and was sold in 50-pound cases for about $12. "The powder is packed in sticks in heavy waxed paper. A few of the sticks could cause the explosion," an executive of a Vancouver chemical firm confirmed.

But regardless of the ingredients, the explosion could not have occurred accidentally. A blasting cap would be required to detonate the powder; the cap itself could be set off by an electric spark or with an ordinary fuse.

By September 1965 the investigation team was still unable to determine the cause of the mid-air explosion of CF-CUQ. "How can I be expected to know, then, just what did cause that crash?" asked Coroner Glen McDonald at a convention of First Aid Attendants on September

24. “Determining cause of death is one of the most difficult tasks facing a coroner or pathologist.”

When a United Airlines Boeing 727 crashed on August 16, 1965, near Lake Forest, Illinois, killing 30 persons, a bomb was again suspected. McDonald joined 150 other coroners at the National Association of Coroners meeting in Milwaukee to discuss and compare this crash with that of Canadian Pacific Air Lines DC-6B. Topics discussed included examination of motives for aircraft sabotage, bomb detection procedures, and ways to tell whether secondary explosions contributed to the crashes. Grisly topics indeed – with open-ended beginnings but disastrous and often unsolved conclusions.

The Lady in the Lake

Through a water-related twist of fate, a Canadian who'd been a fugitive for 24 years was identified and arrested on August 30, 2006 at his home in the city of Plano, Texas.

Acting on a tip following an Internet release of a Kalispell, Montana *Daily Inter Lake News* article, a woman phoned the Flathead County Sheriff's office in Kalispell advising them to search in Plano at a particular address in the Russell Creek Estates area for a man by the name of Michael Lee Smith. His real name was Jarislaw (called Jerry or Gary) Ambrozuk.

It was the culmination of a long flight from justice, and it came about thanks to an Internet advertisement Ambrozuk, a.k.a. Michael Smith, had placed – seeking a new girlfriend.

On August 22, 1982, Jerry Ambrozuk, age 19, from Burnaby, B.C., rented a Cessna 150L (C-GICK) from Sol Air Aviation in Richmond, where he had taken his flight training. David Oliver of Sol Air was of the understanding that Ambrozuk wanted to log a few hours of local practice flight. Further, the plane was not equipped with an Emergency Locator Transmitter [ELT] as required by law for flights more than 25 nautical miles from base.

Ambrozuk took along his girlfriend, Diane Babcock, 18, and against regulations flew over the mountains to Penticton. From there Diane phoned her parents saying they were just getting set to fly back to Vancouver.

Ambrozuk's return flight plan was extremely detailed. From Penticton he planned to follow the highway to Princeton and Hope, the Fraser River to Spuzzum, over to Harrison Lake and Harrison Hot Springs, and again follow the highway to Pitt Meadows and on to Vancouver.

They took off from Penticton at 6:30 pm – and at 10:00 pm the aircraft was reported missing.

After searching for a week, Search Captain Robert DeKelver noted to Vancouver *Sun* reporters that they had discovered "no clues and no traces."

At noon on Monday, August 30, the phone rang in the Vancouver home of a friend of Jerry Ambrozuk's (publicly unidentified, but said to be a student at Simon Fraser University). It was Ambrozuk, calling collect, saying he was in New York. The friend expressed disbelief. "Actually, I'm in a small town," Ambrozuk said, and hung up, leaving his exact location a mystery.

On Tuesday, August 31, a second call came, and this time Jerry's father, Ted, was also present in the friend's home. Jerry said he had crashed the plane in a small lake, and had managed to get out but Diane had not. He'd swum ashore and spent two days in the bush, too upset to call anyone or try to come home. He identified the lake as Bitterroot, near a town starting with the word "White", and hung up.

Both the friend and Ted Ambrozuk positively identified the voice as Jerry's, and the call was reported to police and traced to a bus depot in New York City. The search for the aircraft was redirected 250 miles in the opposite direction to the flight plan, southeast to Little Bitterroot Lake, Montana, near the communities of Whitefish and Kalispell. When no sign of the aircraft could be seen from the air, U.S. state officials began to make sonar sweeps of the 230-foot-deep lake, and to question people in the area who might have seen or heard unusual noises. One lakeside resident recalled hearing a splash about 11:00 pm on August 22, but hadn't checked it out.

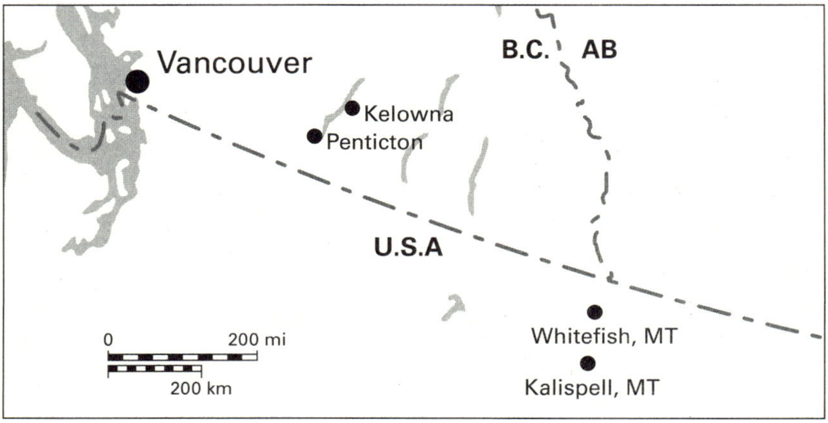

The Flight to Little Bitterroot Lake

But campers had noticed a man who resembled photos of Jerry Ambrozuk sitting at a campfire near the lakeshore that night. His clothing had been soaking wet, and he carried a dry duffel bag but had no camping gear. Investigation of the campsite uncovered burned-out aircraft radio equipment (later confirmed to be from a Cessna 150) in a barbecue pit at the edge of the lake. The equipment included the plane's microphone, which had an on-off switch for transmission. An investigator with the U.S. Federal Department of Communication confirmed that, "If someone wanted to ensure that a radio would not be inadvertently turned on – possibly attracting searchers – they would remove the switch and damage the radio."

Also nearby was an opened green plastic garbage bag that had been sealed with black tape, and had a tow-rope attached, which had probably contained the dry duffel bag.

Back home, police officers and newspaper reporters interviewed friends and families of the two missing persons. Both Ambrozuk and Babcock were popular students and outstanding athletes (he in football, she in track) at John Oliver Secondary School in Burnaby. They were described as being mature and responsible, and had been going together for the past 15 months. Jerry had obtained his private pilot's licence the previous December and had also completed a course in mountain flying, with plans to take geography courses at Simon Fraser University in preparation for a career as a helicopter pilot. Diane had begun studies at British Columbia Institute of Technology in the psychiatric nursing program.

It was then revealed that Diane had withdrawn her savings from the bank the day before their flight, and Jerry had more than $1,000 with him at the time of his disappearance. Was it possible that they'd planned to elope, and things had gone terribly wrong?

"It could be that he is totally confused and in fear, and that he doesn't want to come home, but he wants to reassure his family that he's still alive," was the conclusion of police and family members in reference to the mysterious phone calls. With regard to the elopement theory, Diane's father acknowledged that they seemed serious about each other but had never spoken of marriage. "She went through the typical problems that teenage girls have, but all in all was very emotionally stable and a hard worker … and that just adds to the bizarreness of the whole situation," he said, adding that Diane was more interested in pursuing her studies than an early marriage.

On Thursday, September 9, 1982, Ambrozuk's friend received a third collect call, which was recorded by the RCMP. Ambrozuk repeated what he'd said earlier, that Diane had died in the aircraft but he'd managed to escape and swim to shore. He seemed upset that the aircraft had not yet been found. Then, in a rambling monologue that was transcribed as a 71-page document, he stated, "I went because I was going, and she went because I guess she was in love with me or something like that. She tagged along ... I didn't kill her or nothing."

Why don't you come back, Jerry," his friend asked. "They're looking for you all over the place … . They're gonna look for you, you know, forever."

"... and they'll never find me," Ambrozuk said.

He then gave more specific directions for a search. He was flying on instruments to Montana on a "pitch dark" night, dodging canyons, and didn't see the lake until he'd hit the water and the aircraft flipped and began to sink. He got free of his seatbelt in seconds but couldn't quite recall how he'd gotten himself out of the aircraft. Diane, thrown upside down in her seat, "said she couldn't get her seat belt undone," and he could not save her before the aircraft sank. He did not explain how he'd managed to get all the money and his baggage out. This call was traced to Dallas, Texas.

Offshore Engineering Corporation of Vancouver offered their Deep Access Reconnaissance Television [DART] equipment to probe any target in the lake that was spotted by the sonar scan. This remote-controlled robot viewer could supply detailed underwater pictures and, with the attachment of auxiliary equipment, recover the wreckage.

Now, using Ambrozuk's more detailed directions, Canadian Lake and Ocean Salvage Team Ltd. and Montana police searched through 13 feet of fine silt at the base of Little Bitterroot Lake. Finally, on September 15, they located a metal object, and raised it from the depths.

The discovery of a Cessna 150 with a body inside with long blonde hair trailing out the door and wearing a striped shirt similar to that worn by Diane Babcock, brought emotions of shock and sorrow, and also increased the macabre aspect of this event. On Friday, September 17, police identified the body found in the aircraft as that of Diane Babcock.

Margo Harper, the Vancouver *Sun* reporter who covered much of the on-site investigation, relates that "the case still haunts her."

It was thought that Ambrozuk had approached Little Bitterroot Lake from the north around 11:00 pm the night of August 22, perhaps intending somehow to set the wheel-equipped plane down on the lake. Montana Sheriff Al Grierson voiced suspicion that the Cessna did not *crash* into the lake, but rather the pilot had cut power to the engine in an attempt to land in the water without floats.

Judging from Ambrozuk's call to his friend, and expert opinions, it is thought that he misjudged his angle of descent and his altitude. The right wing might have hit the glassy water, causing the plane to sink immediately. U.S. Air Safety Investigator Gregory Feith added that, due to water pressure, the pilot would have had to open the doors immediately, before the plane began sinking, in order to make his escape.*

---

* Aircraft with fixed undercarriages don't skim the water like a speedboat, but rather dig in and sud-

Once the plane had been lifted out of the water, emergency workers encountered no problems with the release on Diane's seat belt. Her shoulder strap was found unfastened, and the seat belt fastened but not jammed. Had neither Diane nor Jerry been able to release Diane's seat belt, so he was forced to grab his baggage and leave her to her fate?

Inside the aircraft was found a similarly sealed green plastic garbage bag containing women's clothing, black hair dye and a home permanent kit, which perhaps indicated Diane had planned to colour and perm her blonde hair. They also found an inflatable rubber raft. The aircraft logbook contained no details of Ambrozuk's flight from Vancouver to Penticton, or the planned return.

No marks were found on Diane's body, but Ambrozuk might have suffered an injury from the crash – four people who identified Ambrozuk as the person at the campfire noticed he seemed to be in pain and was favouring his left shoulder, perhaps the reason he'd been unable to free Diane, who he said was unconscious after the crash into the lake.*

A warrant was issued for Jerry Ambrozuk's arrest, charging him (initially) with illegal entry into the United States, negligent homicide, and a felony theft of an aircraft.

The mystery deepened further when it was discovered that, several weeks before the crash, the couple had inquired at Kalispell's Frontier Airlines office about the cost and route for commercial flights from Kalispell to Penticton. They were told of two daily flights from Kalispell to Denver, and from there via Vancouver or Calgary to Penticton. After being quoted the fares, they'd thanked the agent and made no more enquiries.

Little Bitterroot Lake residents were not happy with the publicity the incident brought to their pristine environment. "How dare he land in our lake?" reflected the general sentiment. But the local constabulary worked diligently to discover the reasons behind this misadventure, and to try to find Jerry Ambrozuk.

"If you're a cop, you've got to be realistic," said Flathead County Sheriff Al Grierson. Ambrozuk's description was entered in the U.S. National Crime computer and forwarded to Interpol, with a special request that the information also be sent to Poland, Jerry's parents' home before they immigrated to Canada.

On September 21, 1982, Diane Babcock's body was buried in her grandparents' hometown of Trail, B.C.

---

denly stop, pitching nose down in the process. The occupants are frequently knocked unconscious, especially without good shoulder restraints. Simulated and actual ditchings have also determined that the aircraft usually must sink and the cabin fill with water sufficiently to equalize pressure before the doors can be opened.

* In a later statement, Ambrozuk provided a very different description of the situation, saying that "he heard Diane yelling that she could not get her seatbelt off."

Meanwhile, the police in Dallas continued to follow up leads emanating from the source of Ambrozuk's last phone call. It had been placed from one of five phones near a 24-hour Shepp's Food Store at the east side of downtown Dallas, but no one in the area who was shown his photo recognized him. It was believed he'd escaped to Mexico and was using the alias of Luis Gomez, which his dark hair, eyes, and complexion might help to support. It was also reported that he spoke Spanish.

Fleeing pilot felt 'like a murderer'

24 YEARS LATER: B.C. man found living in U.S. after plane crash killed his girlfriend

STORIES BY SALIM JIWA
STAFF REPORTER

A call out of the blue has ended the mystery of Gary Ambrozuk of Burnaby, who vanished after a plane crash 24 years ago in which his 18-year-old girlfriend died.

Ambrozuk, now 43 and living a wealthy lifestyle under an assumed name, was arrested in Plano, Tex., on Wednesday after a woman told a sheriff in Montana that she knew the fugitive.

He is charged with negligent homicide in the death of Dianne Babcock of Vancouver. She was found drowned inside the plane Ambrozuk had rented in Penticton and ditched in Little Bitterroot Lake in Montana on Aug. 22, 1982.

Authorities suspect he deliberately crashed the plane in an elaborate scheme to elope with Babcock.

He swam to safety but Babcock, who had put some of her clothing in waterproof bags and had withdrawn all her savings, died.

Sheriff Jim Dupont of Kalispell, Mont., one of the original investigators, said Ambrozuk, a three-time target of the TV show *America's Most Wanted*, was living under the name Michael Lee Smith in an expensive home in a wealthy area of Plano. He drove a posh Dodge Viper, which costs about $100,000.

Ambrozuk, 19 at the time the Cessna 150 went down, had filed a flight plan for a trip from Penticton to Vancouver. Instead, he veered south, and dived into the lake 100 kilometres south of the border.

Police eventually found the sunken aircraft with Babcock inside, still strapped into the seat.

Her long blond hair spilled from a cockpit window as the plane was raised. In it, they found a hair-dyeing kit that suggested she was planning to alter her identity.

A salvage crew (above) pulls a single-engine plane — containing the body of drowned 18-year-old Dianne Babcock (top right) — about a month after the aircraft crashed in a small Montana lake on Aug. 22, 1982, with her boyfriend Jerry Ambrozuk (bottom right, in 1982) at the controls. Twenty-four years later, Ambrozuk, originally from Burnaby, has been arrested in Dallas on a warrant for negligent homicide. — CANWEST NEWS SERVICE

Ambrozuk apparently spent the night in front of a campfire. Police found charred remnants of flight communication equipment in the ashes. They surmised the equipment had been removed from the plane to prevent any signal from helping pinpoint the crash site.

Dupont said cash and other items were retrieved from the plane before it sank.

Ambrozuk fled without contacting authorities and ended up days later in New York.

Shortly afterwards, he phoned a friend in B.C. and told him he was unable to free Babcock because her seatbelt was tangled. But Dupont noted the seatbelt was not jammed.

"I tried to get her out," Ambrozuk reportedly said in the phone call. "I feel like a murderer. I'm so depressed."

Dupont's tip came from a woman who saw a story about the mystery on a newspaper's Internet site.

"I wish I could talk to Ambrozuk and find out what happened during those few minutes in the water," Dupont said in the story. "I'd still like to talk to that SOB. She was a beautiful girl. She didn't deserve to die like that."

Ambrozuk's father, Ted, a retired Burnaby mechanic, shrieked when he learned his son is alive.

"Where is he? What name is he using?" he asked.

The last time he heard from his son was a Christmas card sent from the U.S. the year he disappeared.

"It is a big 24 years, you know," he said. "I figure he thought he had made a mistake, maybe he was ashamed, maybe he was scared and I think that was in his mind, but in my mind I prayed that he would come back."

Ambrozuk's sister, Renate Wierzbicki, 38, said the past 24 years "have been hell" for her parents.

"It's one thing to grieve over someone's death," she said, but "the unknown" her family has lived with brought its own kind of anguish.

Babcock's grandmother, Dorothy Babcock, said from her home in Trail: "It was such a sad, sad thing.

"She had graduated and had put money down to go and become a nurse. She didn't know she was going to die."

sjiwa@png.canwest.com
— with News Services

Four months shy of retirement, sheriff solves nagging case

Sheriff Jim Dupont hopes to retire after solving his last open case.

HAUNTING MYSTERY:
Texas woman calls in case-breaking tip after man opens up

Flathead County Sheriff Jim Dupont had one nagging file that he desperately wanted to solve before his retirement in four months.

The Kalispell, Mont., sheriff thought it unlikely he would solve the 24-year-old mystery of what happened when a plane rented in

22, 1982, killing a woman.

There were many questions. How was the young pilot able to get out alive? Why did he not stay to answer questions? If he was in love and the pair were eloping, why did he not save her? And where did he go?

The haunting mystery ended for Dupont on Monday when he took a call from a Texas woman.

"On Monday this week I had a person call me that said she thought he was the person we were looking for," said Dupont.

"[The missing man] had alleged-

person then went to the Internet and did a Google search on that name and found my name on a story. So she called me directly and said, 'Please don't think I am crazy but . . .' and she related what she was told.

"Fortunately, she got me because I was the lead investigator at the time and I was a detective at the time and now I am the sheriff. I vividly remember the file.

"I am retiring at the end of the year and this file is the only open case I have. I never thought this [was] going to happen — one

would have wished I had cleared it out.

"I had only four months left [to solve the case.]"

The caller also said the man had been arrested in Dallas on a drunk-driving charge 12 years ago. Dupont phoned Dallas and asked for booking records.

"As soon as they sent us his picture we knew it was him," he said.

"We found him living in Plano, Tex., and he works for a computer company. He is very successful — he is driving a Viper, if that tells you anything, and lives in a very influ-

The Vancouver Province, *September 1, 2006*

On October 13, 1982, the ill-fated Cessna was brought back to Vancouver. It was found to have been in sound mechanical condition prior to its crash into the lake, and it still contained a half-full tank of gas (approximately 20 U.S. gallons), indicating that they must have stopped somewhere en route to refuel (another report, however, says the right tank contained four gallons of fuel and the left tank was filled with water). Following replacement of water-damaged instruments and some structural repairs to the leading edge of the right wing and the engine cowling, the aircraft was tested for fitness to fly again.

On December 20, 2000, the Vancouver *Sun* ran a special series featuring B.C.'s unsolved mysteries. The first was titled "Where is Jerry Ambrozuk?"

The article noted, "Flathead County Corporal Jim Haupt said the crime is the most sensational of the past half-century in northern Montana, and has been featured three times on the television show *America's Most Wanted*. Each airing generates a trickle of tips, but none have led them any closer to Ambrozuk."

Corporal Haupt felt that by this time Ambrozuk could be anywhere – but some day the fugitive's conscience would get the better of him and he would turn himself in. Or, by a fluke, he might get stopped for a traffic infraction and be identified. "When you're not expecting it, something will turn up."

Sun reporter Harold Munro also interviewed Jerry's father, Ted Ambrozuk, who remained hopeful that his son might some day return. "If he's alive, he'll probably show up. I pray for this." The last he'd heard from his son was a postcard sent from Dallas the first Christmas following Jerry's disappearance. The message read, "Only God knows what's going on here. Nobody else."

Solving the Mysteries

Following the arrest in Texas in August 2006 of Jerry Ambrozuk, a.k.a. Michael Lee Smith, many interested parties anxiously awaited the testimony to be given in court. How did Ambrozuk manage to secure a new identity in the U.S.A? How did he accumulate such wealth over the past 24 years that he could afford an expensive house where he'd lived, alone, since the year 2000, and a top-of-the-line vehicle, a Dodge Viper said to be worth over $80,000, parked in his driveway? A local homeowner's directory said that Michael Smith was the owner of AJACOM Corporation, a staffing consulting firm located in the nearby town of Richardson. So, what was his purpose in advertising for a girlfriend on the Internet?

"Now that work is no longer a priority, I do want to meet that someone 'special' to share things with, spend more time with and eventually grow old with," he'd stated in his ad on a Yahoo dating service site. Then he'd added, "A couple of things about me: I am honest and don't cheat or play games. I also believe that there are people that you come across in your life that are very special, but there are very few of them that you can call 'soul-mates.'"

Well, not so honest. Perhaps due to a typo, he stated his age as 34 when he was actually 43, and signed his new name: Michael Lee Smith.

A woman from Texas (who remains anonymous) answered the ad. They began dating and seemed to get along quite well. He revealed to her that his one true love in life had died in a plane crash, and then came to trust his new lady friend to the point where he told her his real name and date of birth. Red flags went up, and she went home and checked out his name on Google. Up came a one-year-old story from the Kalispell, Montana, *Daily Inter Lake* newspaper, displaying a high school photo of Ambrozuk. The lady went to "Michael Lee Smith" with her discovery and suggested he turn himself in. He acted surprised that he was on a wanted list, but did nothing about it.

The lady friend was caught between doing right by reporting her discovery of this long-time fugitive or giving him a break – after all, they were getting along quite famously. Finally, on August 28, 2006, after several weeks of indecision, she placed a call to the Montana police. She sounded extremely nervous as she spoke to Flathead County Sheriff James R. Dupont, giving him the information they needed to find Ambrozuk, a.k.a. Michael Lee Smith. Dupont thanked her and immediately notified the Plano, Texas police.

Sheriff Dupont remembered the case very well. He had been present when the Cessna was recovered from Little Bitterroot Lake 24 years prior, and still retained the image of the pretty, long-haired blond lady trapped in the sunken aircraft. He also remembered that the seatbelt buckle had not been jammed, but sprung open easily when rescuers attempted to bring her out of the aircraft. Later examination showed she had not tried desperately to free herself: no broken fingernails or scratched hands. The autopsy simply showed death by drowning. No money was left in the aircraft. Ambrozuk had taken it all, and run, changing his identity and never contacting family or friends again after the few phone calls.

The police did due diligence, discovering that in Texas in 1983 Ambrozuk had obtained false identity under the name of Michael Lee Smith, complete with a new birth date and a social security number. He had been arrested in 1983 and again in 1984 on charges of burglarizing vehicles, but his false identity had never been discovered. The plan to arrest him in his $300,000 home in a quiet, upscale neighbourhood of Russell Creek Estates in Plano, Texas, was not easy – Ambrozuk refused to answer the door. Finally, a scheme was hatched to catch him for unlawful watering of his lawn, and the arrest was made on August 30, 2006, on a charge of negligent homicide. Smith admitted his true identity and asked for a lawyer. The *America's Most Wanted* program recorded his arrest as AMW Capture #902.

Upon receiving confirmation from Plano authorities that they had their man, CanWest News Service reporters contacted Ambrozuk's family in

Burnaby, B.C. They had not been told about their son's arrest. "Where is he? What name is he using?" Ted Ambrozuk asked. Jerry's sister, Renata, who was just 14 when her brother disappeared, said that family members would fly to Texas to see this brother who had been missing for nearly a quarter-century.

Smith's neighbours in Russell Creek Estates in Plano were surprised that the high-living computer consulting company owner was a fugitive from justice. He'd lived first in New York and then in at least seven different areas in Texas since 1988, from a modest home on Parkview Avenue in Dallas, to Cap Rock Drive in Richardson. Under his new identity he obtained a college degree and then started a software development company. His fortunes increased, and since 2000 he had successfully settled in the usually quiet upscale neighbourhood of Russell Creek Estates in Plano, where people minded their own business and tried to ignore Smith's penchant for hosting "naked pool parties". One of his neighbours said he could see into Smith's back yard from his second-story window and sometimes caught more than he bargained for.

It would prove to be a difficult case: fingerprints could not be obtained from the waterlogged aircraft, and Ambrozuk had never been fingerprinted in Canada. That is why, when arrested under the name of Michael Smith for car burglary in Dallas, the authorities had never been able to connect him with the Little Bitterroot case. Further, his appearance had radically changed in 24 years and even Diane Babcock's sister, who remembered him well when he was dating her sister, could not positively identify him from his arrest photos.

The Courts

Jarislaw Ambrozuk, a.k.a. Michael Lee Smith, at first refused to be extradited to Montana. This left the governor of Montana 90 days to issue a warrant requesting that the fugitive be transferred to Flathead County to face a criminal charge of negligent homicide in the death of Diane Babcock. "We planned to elope and run away from home … . It was a consensual decision we made over a period of months," Ambrozuk said at the bond hearing.

Attending a district court hearing in Texas, Ambrozuk/Smith faced another charge: being a fugitive from justice, which required a posted bail bond of $100,000 before he could be released. However, Homeland Security issued an immigration halt, so even if he posted bail this organization could take custody of him for being illegally in the United States. Bond was denied, and he was extradited to Montana in September, and the court case began.

On Thursday, October 5, 2006, Ambrozuk entered a plea of "not guilty" in district court, Flathead County, Montana, on a charge of negligent homicide. Conviction could result in a 10-year prison term, along with federal charges for having a fraudulent U.S. passport.

Then, on Thursday, February 1, 2007, Ambrozuk appeared before Kalispell, Montana District Judge Stewart Stadler and, in a surprising move, changed his plea to a felony charge of negligent homicide in the death of Diane Babcock from "not guilty" to "no contest". The plea agreement had been reached between Flathead County Attorney Ed Corrigan and Ambrozuk's attorney, Pat Sherlock, and again it was emphasized that this could bring him a prison sentence and a fine up to $50,000. Sitting in the court was Flathead County Sheriff Jim Dupont, who expressed doubt that Ambrozuk was telling the whole truth. "If it happened the way he said it did, then why didn't he come forward?" he asked. "Just landing a plane in the lake at night was negligent."

The deal would also see Ambrozuk plead guilty to felony charges of criminal endangerment and criminal mischief, and federal charges of passport fraud. Flathead County attorney Ed Corrigan's office agreed to drop the additional felony charges of airplane theft and criminal mischief. This allowed the courts to skip the trial phase and move directly to sentencing, which was scheduled for March 8.

Then, in another surprise move, Ambrozuk was allowed to reinstate his original "not guilty" plea on the charge of negligent homicide – the second reversal in the case in the past five weeks. Attorneys Pat Sherlock and Ed Corrigan had agreed to recommend 10 years of unsupervised probation, with Sherlock asking for a deferred sentence that would be removed from his record after probation was successfully completed. Although Ambrozuk would still face federal court action and possible penalties in Texas involving passport fraud, he could now be found innocent, or convicted at trial on the negligent homicide charge, with a new sentence considered. Also to be considered was the court testimony of probation officer Kyle Hinzman that if Ambrozuk were deported to Canada following federal action that he could end up facing no consequences for the negligent homicide charge. The U.S. would lose all jurisdiction at that point, so not even the 10 years' supervised probation could be enforced. On the judge's suggestion, Ambrozuk decided he wanted to stand trial after all.

District Judge Stewart Stadler pointedly asked Ed Corrigan to clarify if what he was hearing from the probation recommendation was "there is not a reason in the world for this court to impose any sentence." Corrigan then reiterated the recommendation: 10 years in the state prison, with all suspended, and asked Stadler to schedule a trial date within the next 80 days. Ambrozuk would remain in jail, where he had spent the last 190 days since his arrest and extradition from Texas.

The Babcock family, father Gerald (whose wife had been killed in a car accident in August 1999 en route to visit their daughter's grave), Diane's sisters Linda Babcock and Jodi Stephen, and brother Geoffrey Babcock, attended the hearing. It was the first time they had ever seen their father cry.

Jerry Ambrozuk's sister, Renata Wierzbicki, spoke on behalf of her brother and their parents, Helena and Tadeusz (Ted) Ambrozuk, recalling the closeness she'd once shared with her "loving, caring" brother who now regretted the foolish actions of his younger years.

It was not hard to believe the suffering felt by both families.

In May 2007, another surprise plea agreement was reached. With Ambrozuk's trial scheduled to begin on Monday, May 9, he pled guilty to criminal endangerment and criminal mischief. Charges of negligent homicide and theft of the Cessna 150 aircraft were dropped.

The prosecutors' recommendation of two 10-year suspended sentences, to be served concurrently, was upheld at court on Monday, May 21, 2007. He would then go back to Texas to plead guilty, and face sentencing, on the federal passport fraud charge.

The restitution hearing on May 31 saw Flathead District Court Judge Stadler order Jarislaw (Jerry) Ambrozuk to be fined $1,000 under the plea agreement; $19,500 in restitution to the owner of the Cessna 150L he'd rented from Sol Air Aviation in Richmond, B.C.; $5,000 in funeral expenses to the Babcock family; and $10,000 in court and prosecution costs.

Ambrozuk appeared contrite in his statement to the court. "Taking the plane, ditching it in the water, was not a responsible thing to do, and I'm truly sorry for … her death … . I also want to say how sorry, truly sorry, I am … to the Babcock family and their loss of their daughter." He then added, "I hope that one day they also will find peace, like I have."

Sheriff Dupont still doesn't buy Ambrozuk's explanations. "It still bothers me that he had an opportunity, not only after the crash but for the next 24 years, to come forward. If it happened the way he said it did, then why didn't he come forward?

"I think he just let her go down."

The Murphys of Musk-Ox Lake

"If you had Neill Murphy with you on a flight, you were assured of coming back alive," states aviation pioneer Chuck MacLaren. Indeed, Neill's unique path from a primitive lifestyle as an Arctic Barren Lands trapper to a world embracing the latest in aviation technology offers a story found in very few 20th century biographies.

Neill Murphy was born in South Fort George (now incorporated into Prince George), B.C. on January 27, 1920 to Matt and Margaret Neill Murphy. While Matt's father was of Irish stock, his maternal grandmother was a full-blood Ojibwa from Manitoulin Island in Lake Huron. Four children were born to Margaret and Matt: three sons: Cyril, called "Bud"; Glenn; and Neill; and one daughter, Mavis.

Matt Murphy was a fur buyer for the international Charles S. Porter Company. When Neill was nine months old the family moved to Peace River, Alberta, where Matt was put in charge of three fur trading posts. The main post was in the town of Peace River, with smaller posts at Keg River and Waterhole.

Neill grew up in the world of trappers and fur buyers. "I hung around dad's office, and watched people bring in their fur. Dad would examine the quality, and quote prices offered for the pelts. Then I'd watch him sort the fur and bale it. Dad spoke Cree and I also picked up a few words – in fact, he gave me a nickel for every word I could remember. At that time the Peace River country was known as one of the best producers of fur in North America, and there was lots of it."

In the winter of 1924-25, following a downturn in the fur business in the Peace area, Matt Murphy went to Artillery Lake north of Fort Reliance in the Northwest Territories to trap fox, wolf, mink, and other furbearing animals. It was not necessary to register a trapping territory – a person simply established a camp. Margaret Murphy and the children stayed behind.

Peace River Pioneers

Meanwhile, life at home in Peace River also provided adventure. "It was still a frontier atmosphere," Neill says. "We were pretty well cut off from the rest of the province. There were no cars except the old Model Ts. When three cars came up from Edmonton in 1940 for a demonstration of some kind, it took them three days to get here."

The mighty Peace River was a focal point, for kids as well as adults. "We were fishing all the time," Neill says. Many newcomers entered Peace River by foot, especially during the Depression and drought years that had so harshly hit the Prairies. "Day after day we saw people coming down the hill, driving a couple of skinny old horses, pulling a wagon, kids staggering along, no shoes on," Neill recalls. "It was really pathetic. They'd go down to the riverbank and camp until one of the boats was going farther north, to LaCrete and Fort Vermilion, and they'd settle there. They were after land with water. We were hardly any better off, we ate a lot of cracked wheat – oatmeal – but I think neither these kids nor their parents had seen even much cracked wheat for a long time."

But the town looked out for such people, and also families whose husbands were gone for long periods of time, such as the Murphys. There were times when Matt Murphy wouldn't make it home for two years. "He had to wait until he'd got enough fur so he could afford to come out, and if he didn't he just had to stay there," Neill says. "At that time he couldn't jump on an airplane to come home."

In 1928, the eldest Murphy son, Bud, joined Matt at Artillery Lake. A year later, Matt and Bud moved farther north to establish their camp at the north end of Musk-Ox Lake on the Back River. In 1932 they moved eight miles farther north to set up their camp on a slope above a small unnamed lake between the junction of the Back and Contwoyto rivers. Now based some 275 miles northeast of Yellowknife, and over 200 miles from the closest settlement of Fort Reliance, it was indeed an isolated existence. Sometimes Matt would be away from camp for two weeks with his dog team while he covered his 65-mile-long trapline.

Neill, meanwhile, was anxious to go north as well. His after-school jobs delivering groceries and looking after a neighbour's wood and water supply seemed dull in comparison to his dad's lifestyle. When he got to Grade 11 he felt he'd had enough.

"I had been after my dad to take me north with him, and I finally ambushed him. I said, 'If you don't take me, I'll go back with George Magrum [one of Matt's fellow trappers].' He looked at me and said, 'Well, rather than that, you'd better come with me.'"

Life on the Line

Although Neill says "It was okay, living in the camp up there through the winter, with dad and my brother," their home place could only be called primitive.

"It started out as a tent, canvas stretched over poles," Neill says. "Then it was lined on the outside with caribou hide, with moss on the roof plus caribou hide, and sided with sod all the way around on a gradual slope to bank it up."

With fuel oil much too expensive to fly in, the hut was heated with a homemade oil-drum stove that burned small scrub willow twigs – "a lot of them!" Neill says. "Sometimes I'd be lucky and get something an inch in diameter and maybe 10 feet long. A lot of the ground birch and willows lay low and when I'd pull on them I might get 10 foot lengths."

He produces a photo of his father standing in the doorway. "Dad was born in 1888, and was 72 when he died in 1960," Neill says. "He always did look young." In front of the hut is a chair that Neill made out of caribou hide and antlers. To the left, a white structure (the toilet) is shrouded with canvas and hides. A hole would be dug into the permafrost, and that sufficed.

"I had my own dog team and Dad had his," Neill says. "I was more or less travelling with my older brother. When we left to go out on our lines my dad would go one way and Bud and I would go another." Neill usually travelled with a five-dog team. If he had a heavier load, perhaps when hauling a lot of meat, he'd add another couple of dogs.

Life at Musk-Ox Lake was busy. "We had a bunch of dogs to look after and feed, and in the summertime make sure they had water. We had a fishnet, and fished in the summer and through the ice in winter. In the fall we'd hunt; there were lots of caribou.

"We had to take care of our fur – skin, flesh and stretch it, and we'd tan caribou hides. We made our own pants and parkas, sewing them ourselves with needle and thread, and our own footgear made with moose hide and stuff we'd brought from outside. It was good."

Bud left the line in 1939. "He came down to Reliance, left the dogs there, got on an airplane that happened to be in there, and flew to Yellowknife. He then came back to Peace River and got married."

At the time Bud left, Neill had been outside for three weeks. News of the war had penetrated the far north, and Neill went to Edmonton to enlist.

"First, this fellow asked me where I was from and I told him I was from the north, out on the barrens just south of the Arctic Circle. He said, 'You'd better go back and just keep your eye open for possible invasions.' I laughed and said, 'We have no communication, and a dog team is very slow in comparison with radio!' I didn't actually catch what he was getting at, but when I went to take the medical exam for recruitment, the doctor didn't pass me."

It was then that Neill discovered the name for a physical anomaly he'd had all his life: Tourette Syndrome. His affliction was fairly mild, confined to uncontrollable motor tics resulting in rapid eye blinks and occasional face grimaces. But the malady was there nevertheless, and was the reason for his rejection for military service.

When he returned to Yellowknife he was hoping to catch a flight back to Musk-Ox Lake. However, no one was flying until the spring thaw started, when pilots could spot rocks and other landmarks beneath the snow. Neill stayed in Yellowknife for a month and finally caught a flight with pilot Mickey Found (who later, with his brother, established the Found Brothers Aircraft Company). It was his first flight in an aircraft.

The Legend and the Legacy

Matt Murphy was renowned as a true man of the north, who could survive for an endless period of time with just the bare necessities. But Neill recalls an incident in 1941 when even Matt got tired of an unending diet of "midwinter wilderness provisions."

They'd had a good catch of fur and wanted to stay until the end of the season, the first of March. When the RCMP arrived on their regular patrol to check on the various trappers, Matt and Neill asked them to arrange for an airplane to pick them up. Their alternate choice would be to travel south to Fort Reliance with their dog teams, which would take five or more days, and from there another two-day journey to Fort Resolution. If an aircraft could come in toward the end of February, they, their furs and their dogs could quickly be transported out to Yellowknife.

"We had been running short of grub – we had some tea and sugar and that's about all, and of course lots of meat – but anyway the airplane didn't come and didn't come," Neill says. In fact, their flight was scheduled near the time when aircraft were going down to Cooking Lake to change over from skis to floats and then coming back, so there was a time lapse of possibly three weeks when there were no flights out of Yellowknife.

"We'd pulled up all our traps and got everything straightened around, so there wasn't a heck of a lot to do but lie on our bunks waiting for the airplane. Finally Dad said, 'If they don't show up by the end of the week we'll knock these two toboggans apart and make a sleigh, load everything on that, and away we'll go.'"

One day Matt was bemoaning their constant diet of meat and fish, when Neill remembered where there might be some leftover rice. A couple of years previous they'd cooked some for the puppies, and he had buried a leftover can of rice in a little dugout behind the shack.

"So I went out and sure enough, this can was still there. I took it in the house and opened it up. It was just a mass of mould, but just for the heck of it I washed it all off and got it halfway like it was white rice, and decided to cook it up. Dad and I looked at each other and at the rice, wondering who was going to be the first to try it."

While they were pondering their choice of whether to risk food poisoning and eat it, or throw it out, they heard a sound.

"We'd got an old aircraft control cable from a wreck on Artillery Lake, and had it running from the peak of the roof at the back down to an anchor to help stabilize the camp in the high winds," Neill says. "Every once in a while when the temperature changed this thing would hum. Just at that moment we heard a humming sound. 'We must be getting some warm weather,' Dad said, but I said, 'No, that's an airplane!'

"I ran to the door and he was right behind me. But as Dad ran out, he grabbed this pot off the stove and just heaved it over the bank! Luckily I was right: it was an airplane, so we didn't have to eat mouldy rice."

Their saviour was pilot Alf Caywood arriving in a Norseman on the last flight of the season. Caywood had once prospected in the north; after becoming a licensed pilot he filed sketch maps of locations of the various outlying prospectors and trappers. Now, flying for Canadian Pacific Air Lines, he and the fabric-covered single engine Norseman were a welcome sight.

Neill and Matt stayed overnight in Yellowknife, found a place to keep their dogs, and chartered a Canadian Airways airplane to go from Yellowknife to Edmonton. "We went to Hay River, took the skis off, then to Peace River, which was our real destination, but we weren't allowed to deplane. There was a DOT [Department of Transport] regulation that stated no one could take passengers unless the airplane was chartered, and no one could get off in the middle of a chartered trip. Because we were in a Canadian Airways aircraft and on a Mackenzie Airways route, we couldn't even offload our stuff, which was really no problem as the furs had to go to Edmonton anyway. Then we came back to Peace River from Edmonton on the train."

The Trapper Takes a Wife

In 1941, Neill met a pretty young girl in Peace River by the name of Jacqueline Vowel, a local farmer's daughter. They were married the next year, when Neill was 22 years old. The farm girl wasn't afraid to try wilderness living, and happily went north to stay at Musk-Ox Lake with her husband and father-in-law. The 10-by-12-foot hut was heightened by digging 18 inches deeper into the permafrost, and enlarged to a length of 30 feet by adding a back bedroom to accommodate the newlyweds.

"There was no reason to build another camp," Neill says. "You see, when you're trapping you don't just leave in the morning and come back at night. I might be gone a week, and then Dad might leave the day after I got back and be gone for a week, so we wouldn't see each other for maybe a couple of weeks."

Jackie – or "Johnnie" as Neill called her – often accompanied Neill when he went out to check the line. They stayed in tents that Neill had set up at intervals along his trapline. The 7-foot-by-7-foot tents were secured by bent willow boughs and tethered to small adjacent bushes by ropes, or preferably haywire, which the Arctic hares couldn't chew through. Heat was provided by a homemade stove fashioned from four-gallon gas cans, and vented with stovepipes protruding through the roof.

And Baby Makes Four

Jackie went out to Peace River for the birth of their first child, Terry, on August 16, 1944. Seven months later she returned with Terry, and the four Murphys and 16 dogs continued to live happily in the wilderness. Jackie had her own dog team – Neill says "she was good at driving" – and ran her own trapline.

When Don Cameron Exploration left a shortwave radio ("5 Roger") plus the necessary engine and generator so the Murphys could contact Yellowknife and provide weekly weather reports, they were suddenly connected with the outside world. "The reception was exceptionally good. We could hear broadcasts from anywhere in the world," Neill says. "The standard band was a little tricky because the Northern Lights would often cause it to fade out, but shortwave stayed clear."

Although months might pass before the Murphys actually saw another human being (their nearest white neighbour, Sam Otto, was 30 miles away), the radio allowed them to send and hear messages over the CBC's Northern Messenger Service broadcast every Friday night. Everyone in the north tuned in to get the news. With the radio, the Murphys could also be patched through for calls to far-flung family, friends and neighbours.

One day an aircraft arrived to pick up a radio that had been left there the previous fall. When a fellow on board saw that Neill had his dogs hooked up ready to go for a load of willows, he asked to come for a ride. As they travelled over the tundra they came upon a cluster of frozen yellow columns sticking two feet high up from the snow.

"What on earth are those?" the passenger asked.

"Some caribou stopped here to have a pee," Neill replied.

"Jesus Christ it must have been cold!" the man exclaimed.

Neill explained that earlier in the winter a small herd of caribou had crossed over a large snow drift, and several of the animals had stopped to urinate. When another storm blew the drift away, the frozen yellow "stalagmites" became a northern anomaly formed by nature's call.

In the fall of 1946 Neill transported a spruce tree from Fort Reliance and planted it beside the hut. To celebrate Terry's second Christmas the tree was brought inside and decorated. It provided an exciting image for Christmas in an almost treeless land.

For a special treat that year Neill bought a 10-pound box of almonds and had stored it, along with other supplies, in a tent 50 feet behind the hut. Just before Christmas he reached under his bed in the hut to retrieve his winter moccasins. A pile of almonds tumbled out! Totally puzzled, he went back outside to the storage tent. The almonds were gone. The only possible conclusion was that, one by one, the mice had transported the almonds from the outside storage tent, and scurried through the snow to burrow through the sod sidewall and store their booty in a nice warm place for their own Christmas treat. Neill laughs as he tells the story. "We just washed them off and ate them."

Jackie's Journals

Jackie adapted quite well to living in the north. The cold didn't bother her – even the coldest day that she recorded at -73° F – although the incessant wind got her down at times. Baking, sewing, cleaning the house, bringing in wood and water, reading, writing daily entries in her journal, and looking after her own trapline, the new pups, and "her men" – Matt, Neill and little Terry – occupied her days.

On New Year's Day, 1948, their bitch, Sugar, gave birth to five puppies. Jackie records the momentous event in her journal:

> *Terry of course was very interested and for the first time became acquainted with the nature and birth of all things. It was a big and delicate moment and we tried to answer all his questions in the most natural manner possible.*

On January 3 Matt left with his dog team to go over his line. On his return, the ground drift was so thick that his lead dog, Rowdy, got off the trail and became lost. Jackie writes:

Rowdy got panicky and started running first one way, then turning directly back and running the other way. Dad finally stopped him, went up and patted his head, which seemed to reassure the dog for he calmed down and went in the direction he was directed. Noting a marker after perhaps a half-hour, Dad knew they were once more on the trail.

Such were the challenges and hazards of running a remote trapline.

Aircraft going overhead were always noted with great excitement. Sometimes calls could be picked up on shortwave, and the Murphys could clearly hear the pilots talking to Cambridge Bay. Scheduled aircraft deliveries and pickups for Musk-Ox Lake were periodically arranged, but at times a "sked" had to be aborted when the pilot considered it too long a trip from Yellowknife for a Tiger Moth in the dead of winter.

On January 5, 1948, pilot Stan McMillan came in with an Anson, bringing food supplies ranging from onions to eggs, and also Christmas letters and treats, oranges and ice cream, rum and chocolates, cigars, nuts and candy. Both the famous pilot and the supplies were warmly welcomed.*

On January 7, Neill, Terry and Jackie went by dog team to check her trapline.

We took an eiderdown along so Terry and I kept warm by covering up our heads and both going to sleep for a ways. Neill froze his face in several spots. There was a bitter wind and it was 38° below, hard pulling for the dogs. After Terry fell asleep I could smoke. He is always throwing his head back and either burning the fur on his parka or bending my cigarettes!

While Jackie generally embraced the northern lifestyle, at times despondency crept in, most often following long periods of extreme cold, winter darkness, and never-ending wind. *It seems as though I long more every day for all of us to be outside, once and for all,* she writes on January 7. On January 18 the thermometer read -54° F, but Jackie was heartened by a quote in a magazine, *No one is bored who is trying to create something beautiful or to learn something new.* And so she concentrated on the beauty of the blue-white snowdrifts reflected in the midday sun, the fabulous sunsets followed by *lots of beautiful moonlit nights, and stars for company* and the wonder of the Northern Lights.

January 27: *We celebrated Neill's 28th birthday; the temperature rose to 10° above zero; Matt got three wolves and Neill one.*

---

* In 1929 Stan McMillan was one of the first airmen to penetrate the Barren Lands, on a record 4,000-mile trip from Winnipeg to north of the Arctic Circle as part of the famed MacAlpine Expedition. Stan continued to work as a northern pilot with Mackenzie Air Service Ltd., Canadian Airways Ltd. and later Associated Airways Ltd.

February 8 was exceptionally cold and so stormy, with a 50 mph wind, that the snow house they'd built for Sugar and her puppies blew away in the wind. When Sugar's scratching at the door alerted them, Neill went outside to dig around in the swirling snow for Trish and Herbie, two of the missing pups.

He looked everywhere, and finally heard a little squall. Trish was sitting in the open by the far radio pole, her face caked with ice and her undercarriage frozen. She was almost played out.

Dad had gone out to shovel the three-foot drift from in front of the door, and Neill was looking everywhere for Herbie with a flashlight and a spade. He finally had to come in to get warm. The wind was terrific and the stinging snow all but blinding. We kept looking without results for poor little Herbie and finally had to give up about 10:00.

We went to bed about 12:00, everybody feeling bad about Herbie and giving up hope of ever finding him alive. But we couldn't sleep for worrying. Finally Neill couldn't stand it any longer. He got up and dressed and went out to look some more. I got up and lit the fire. I could see his flashlight bobbing around through the window.

At 2:30 a.m., Neill came up carrying Herbie! He had found him, howling for all he was worth, down below on the path that leads to the lake. He must have found some sort of shelter because he was warm. He started running around, perky as anything, shaking big lumps of snow off on the floor. We are all so dog-gone glad to have him found. I wanted to hold him and hug him. I think Neilly is wonderful. I did before but more so now. He is so glad too. Finally got to bed at 3:30. Neill froze his face rather badly.

The next day Neill and Matt went to work, shovelling a hard-packed, five-foot-high by 20-foot-long snowdrift that blocked the door. Only Terry seemed to enjoy the winter scene, happily sliding down the hardened banks.

When living in such isolation, good health was of prime importance. "Although it might seem strange, if you're out in a place like that you don't get sick," Neill says. "There was no one to catch anything from. The only medicines we kept in the house were some aspirins and a few bandages."

A visit on February 21, 1948, from a nomadic Inuit (then called Eskimo) family was a welcome occasion.

We were getting ready to go out to check some traps, when Neill spotted, through field glasses, six wolves running along across the river. Then he spotted four dog teams coming down the draw. It was Eskimos coming to visit! They arrived shortly and pulled up at the bottom of the hill. We went down to meet them, and they all came up.

Although they spoke little or no English, "they were a nice bunch," Neill says. "They drank tea, made a couple of snow houses for themselves, and stayed a while. We got along fine."

Jackie writes:

There are five of them including a woman and a boy about seven or so. They all came in and shook hands. I made several pots of tea and we all looked at each other and grinned. Only one speaks a very little English. Terry showed the little boy all his toys. They got a kick out of the collapsible dog. Their names are William, his wife Martha, sons Billy and David, and another David. Little David smiles all the time, but they all do. Boiled a big pot of meat for supper. Boy can they eat! They have some lovely parkas and mukluks.

They built a snow house this afternoon while we watched and took pictures. More tea before bed at 7:30.

The next day the two women went hunting ptarmigan. Jackie was amused to see that Martha carried her own cigarette lighter and tucked it in her garter when not in use.

On February 23, following the two-day visit, the Inuit family *hooked up and left. They waved all the way across the lake and soon were out of sight. We could hear them yelling at their dogs long after. We enjoyed our visitors very much.* Jackie's journal entries cover several pages detailing their dress and mannerisms, which over the years and with radical cultural changes provide fascinating historical insights.

Leaving the Line

"The Ides of March" brought serious discussions between Neill and Jackie about "leaving the line." Terry would soon be ready for school, and a second baby was expected in October.

Murph and I decided that we wouldn't come back next winter so have been making plans. I told Dad today and he seemed glad that we weren't coming back. I am too. When sometimes I wish I was back here again, I'll just remember the drifting snow and the cold and the solitude … .

Jackie's excitement grew as she planned the next phase of their lives.

[I long for] green things and fresh fruit. That is what I propose to buy in Yellowknife instead of rum! … We spend a lot of time making plans these

days. We may mush out to Yellowknife with the dogs. I think it would be fun if it isn't too cold.

They received further visits from an Inuit group, this time five men (Ike, Moses, Daniel, Peter and little Moses), so out came the meat pot once again. Jackie silently watched, and later recorded the mannerisms of these guests:

They fished each chunk of meat out of the pot with their fingers, took hold of it with their teeth, cut off each bite with the knife and chewed and smacked loudly. Then they played Chinese Checkers, and each tried a cigar, for the first time apparently. They had a few choking puffs and threw them in the fire.

By early March, the daylight hours lasted until 6:30 pm.
March 9:

I am truly looking forward to going out and getting settled somewhere. Dad seems to think I'll wish myself right back here but I don't think it will be for some time. We all definitely need a change and many discouraging years in a row do not help matters.

March 11:

The wolves are raising merry hell with the traps. They scratch all the bait from around them and quite often spring them without getting caught. They seem awfully wise in this respect. I guess they can smell the steel traps. However sometimes we are lucky. It has got way past the point where I am sorry to see an animal caught in a trap like that. It's either them or us, it seems.

March 16:

Every time I think about how short the time is getting before we leave, I get excited. It will probably be disappointing in lots of ways but then it is such a different kind of a life than we lead here.

The continuing cold (-38° F on March 17) added to the difficulty of daily tasks.

It is quite a chore digging out the water hole these days and Neill threatens to start using ice every day, but that is an awful job too. The water is full of little wiggly bugs and has to be strained before using. Terry spends a lot of time watching them hop around in the water pail.

Perhaps the news of the Murphys' hospitality had travelled by "moccasin telegraph", because on March 25 they received a visit from a group of 11 Inuit visitors. The snow house was too small to accommodate them all, so six people, including a toddler and a baby, slept on the floor of the hut. *This thing has turned into a nightmare – or a rank comedy, it's hard to decide,* Jackie writes. By the following day a second snow house had been built and the family moved outside.

An order for fresh supplies was booked with Associated Airways out of Yellowknife for April 15 and right on schedule pilot Doug Ireland flew in bringing fruit, eggs, the mail, and a book titled *Wilderness Wife* for Jackie – who by this time could have written it herself. But packing to leave was now the main task.

April 16:

> *Neill brought in seven dunnage bags and eight gunny sacks of junk, and I got the job of sorting and burning – mostly burning. The stove was red hot for a couple of hours and everybody stewed and fretted. This getting ready to go away for good is a big job. There's so much junk collected in our four short years here.*
>
> *I often wonder if we'll ever come back, either out of necessity or by choice.*

But as the temperature continued to hover below zero and storms whipped snow around the hut, spring seemed forever out of reach. On April 22 the Don Cameron Exploration aircraft came in to take back the radio "5 Roger". It had been a godsend, bringing the outside world into their little hut, sometimes through static and crackle but alive with words and images of another lifestyle, which they would soon be sharing.

The next day, Neill took the dogs out west to clean up Jackie's trapline.

> *I almost ran out and told Neill to wait for me but didn't. This is the last trip to pick up traps on my line. Neill sprung them all long ago. Going to miss that line with its three lakes and long winding rocky hill! There's one place coming down a steep hill that is particularly rough. The dogs tear down it like mad and every jump jars your spinal system loose. Terry stands up and yells gleefully.*

The spring blizzard continued, wood became scarce and the hut was *so damn cold we nearly froze in bed … . The wind seems to come from all directions; the roof is leaking in several spots. Snow piled five feet high in front of the door and right over the roof.* Neill had to crawl out though a small dip in the snow at the top of the door to go for water, although the lake was still covered with three-foot-high drifts. While there he spotted a herd of caribou across the lake, and counted about 200. *Speculation is rife as to how the plane will make out when it lands,* Jackie writes. *Too bad 5 Roger is gone. We could use it now.*

April 30, 1948, carries Jackie's last journal entry, describing a sunny but windy day. Neill and Terry went out for a short trip with the dogs, and Terry delighted in splashing in the puddles.

Spring had finally arrived at the little camp in the Barren Lands, but Neill, Jackie and Terry were gone.

The Wild Goose Chase

From his association with aircraft while on the trapline, booking passenger or cargo trips, Neill had come to know several people in the aviation industry. As soon as he moved to Yellowknife, he went to visit pilot Doug Ireland at Associated Airways. Doug had a trip booked with the twin-engine Anson to take a group of scientists – Peter Scott (the son of Antarctic explorer Sir Robert Falcon Scott), Paul Queneau and Harold Hanson – to seek the location of the Ross's Goose nesting ground. Their destination was Perry River (Kuugjuaq), a waterway emptying into Chester Bay, Queen Maud Gulf, in the Northwest Territories (now Nunavut territory).*

The RCAF agreed to bring the three men and their equipment to Yellowknife in a Dakota, and then fly part of it on to Cambridge Bay. Doug Ireland would take the men and the rest of their cargo from Yellowknife to their destination. He would drop them off at Perry River, fly back to Cambridge Bay to pick up the remainder of their equipment, take it to the scientists' camp, and return to Yellowknife. Doug asked Neill to go along, for several reasons. Scott had heard that Matt Murphy was an amateur ornithologist and wanted to meet him, so their first stopover would be Musk-Ox Lake. Neill's wilderness knowledge and general mechanical ability would also be welcomed on the trip north, as well as his company so Doug wouldn't be flying back alone.

They encountered ice fog over Wolverine Lake, about 100 miles northeast of Yellowknife, and set down to wait for it to clear, taking this opportunity to put skis on the aircraft. Two hours later they reached Matt Murphy's camp.

"When we landed on the small frozen lake beside the camp, the snow was so soft that we just mushed in," Neill says. "We tried to get going a couple of times but the air scoops underneath the cowlings became clogged with the wet snow and stopped the engines." And so they stayed, for three days.

Daily efforts to move the aircraft came to naught, but the northern men were used to waiting out weather. "The thaw was on, so it wasn't too long before the snow dropped," Neill says. "The water usually sits on top of the ice for a while and then gradually seeps and the ice rises up on the buoyancy. To speed it up a little we cut several holes in the ice, which was about five feet thick, one at each end and a couple in the middle, and within minutes of the holes being made the water just rushed in. It took two or three hours before

* A book written by Sir Peter Markham Scott titled *Wild Geese and Eskimos: A Journal of The Perry River Expedition of 1949* offers a detailed documentation of this trip.

the lake was drained of the water on top so it left straight ice."

After leaving Musk-Ox Lake they flew north for 50 miles, viewing migrating caribou herds below. When they encountered a snowstorm at Beechey Lake they had to sit for a few hours until it cleared. A mile-long taxi down the lake to gain speed in the sticky snow finally got them airborne.

Over the Western River going into Bathurst Inlet they ran into an even fiercer snowstorm. "It was kind of ticklish going through there," Neill acknowledges, "because of the height of land where the Back River goes down into Beechy Lake, then takes a sharp angled turn to the east. You go over the height of land, which is fairly high, and then the Western River meanders into Bathurst Inlet itself. It was open, so we were watching – Doug watching out his window and me out mine – like, if he didn't see the river I did, or I didn't and he did."

Peter Scott's description of what they were getting into is even more graphic: cloud that dropped down into a gorge that *looked like a tunnel – and we had to fly into it.* Then they encountered heavy snow.

> *The next half hour was, to me, very frightening. Once Doug missed the bend and swung out. The wall of the gorge loomed up suddenly. Doug decided to make a loop back. We did a tight circle and below in the swirling snowstorm we saw two small flocks of geese .... We had scarcely time to look at them, so desperately were we peering at the banks of the river.*

Visibility was reduced to no more than 100-150 yards as they flew through the 800-yard-wide gorge of the river that twisted beneath them.

Neill takes up the tale. "One place there actually needed attention because we both lost track of the river so we had to make a sharp 180 degree turn. Peter Scott had been looking out the window too, and he tapped Doug on the shoulder and said, 'Thanks, Doug, for giving me another look at that bunch of white-fronted geese down there. I didn't know if you'd seen them.' Well Christ, we weren't thinking of geese I'll tell you. But anyway we picked up the river again and went on and hit Bathurst Inlet."

They remained grounded at Bathurst Inlet for another three days, while making several failed attempts to get to the Perry River area. They would nearly reach the Ellice River when it would get socked in again, so they'd have to turn back. And each time they came back they had to refuel.

"The air force had left some barrels there for this expedition – Scott had arranged it from the year before," Neill says. "They had left this gas in barrels in the sand on the shore of the ocean, but with the slopping water and everything they were now frozen solid with the sand. Then, with the winter's snow and drifts, they were buried eight or ten feet deep into the snow banks. We couldn't even see them, but the Hudson's Bay manager knew where they were, so then we had to work hard to get the gas out, with the old wobble pump and buckets. That operation had to be repeated after each attempt to take off, only to be stopped by the weather and have to come back. We finally got away by going down the inlet and around Whitebear Point into where they wanted to go."

They landed, and dropped off the men and the cargo they'd transported from Yellowknife. But on the way to Cambridge Bay for the rest of their supplies, Doug and Neill again encountered fog and snow showers. Cambridge Bay was supposed to have radio contact but didn't – "we found that out!" – so just by luck they spotted the peak of a tower that protruded into the air. This tower – 550 feet high but barely visible through the fog and snow – gave them a small landmark with which to line up, and safely get down.

The anxious RCAF personnel who met them urged Doug and Neill to get some sleep and something to eat, and then offered to bring the scientists' cargo to the plane and load it for them. "When we came down it was all loaded except for a 12-foot aluminum canoe, so we strapped that on the wing root and tied it down good," Neill says.

The air force tried to discourage them from taking off in such bad weather, however, as visibility was below limits. Doug responded, "We have 50 feet more than when we came in! Why didn't you stop us from coming in?"

They were finally, and reluctantly, given permission to take off from the ice of Cambridge Bay. When they were about 10 miles out, and flying 100 feet off the ice, Doug decided to "get some altitude," but 850 feet was all he could manage because of the load. However, they reached the scientists' camp, unloaded everything, and then drew a map of their location so Jim Bell, the pilot scheduled to come in on floats in July to pick them up, could find them.

Scott and his party gave Doug and Neill the last letters they'd be able to send out for seven weeks, and then watched the red and yellow Anson lift off – more quickly now because of its lightened load. Scott writes:

> *We kept it in sight until it was a speck in the south-western sky, and we talked about how much we owed to Doug, his plane, for the success of the complicated plan for getting us here. We were landed with all our gear and supplies, at the very centre of operations. It seemed almost too good to be true.*

Doug Ireland and Neill Murphy then returned to the camp near Musk-Ox Lake, removed the skis from the aircraft, loaded Matt and the dogs, and headed back to Yellowknife.

That was the first time Neill had anything to do with airplanes in a working sense. Considering the risks encountered on the trip, it could easily have been his last.

Neill soon found a regular job in Yellowknife, installing water and sewer lines for the town. When the company had problems getting the necessary labour force, he contacted local natives who were quickly hired on, and Neill was promoted to foreman. When that job finished, Neill was hired as a water pumphouse operator. For the first time in his life, he received a steady pay cheque.

The family got settled in a "shack" down in the Old School Draw (the only road from Yellowknife town to Con Mine along the lakeshore), Terry started school, and Peter was born on October 2, 1948.

When Doug Ireland quit Associated Airways and started flying for Yellowknife Airways Ltd., based in Edmonton, he suggested that Neill should apply for a job with this company. He'd acquire some on-the-job experience and then later could take some aircraft engineering courses in Edmonton. "The incentive," Neill says, "was that one night he came over with two bottles of over proof rum. In the morning I applied for the job, and got it."

He was first apprenticed to engineer Jack MacDougall, a brother-in-law to Associated Airways' chief pilot Art Bell. "Jack was a very good engineer," Neill says. "There wasn't anything he didn't know about an airplane."

"One time we had to change the prop on a Norseman," Neill says. "It's quite heavy and the nut holding it on has very fine threads, so by the time you lift it up and get it onto the shaft – and sometimes those threads don't catch – you're played out. Also, the aircraft was on floats so we had to stand on boards to try to get the propeller installed. We were having a lot of problems, so Jack finally said, 'To heck with it. Put it back down. We'll try a little trick I know.'

"We put the prop down, went into the office and sat around while he smoked a couple of cigarettes. I was waiting for this 'trick', but he never said a word. Finally he got up. 'Okay, let's go.' We went out, picked up the prop and put it on the airplane. The threads caught at the first turn. I said, 'Was that the trick?' He replied, 'Actually, the trick was to ignore the son of a bitch for a while!'

"That was a valuable lesson that I referred to throughout my career," Neill says with a smile. "Go away, ignore the problem for a while, and when you go back the thing will usually work. Don't stay at it until you're so frustrated that nothing works!"

Some pilots and engineers became known for a special constant companion they took along on their trips. Northern pilot Jim McAvoy had his poodle Salty (actually, a succession of Saltys). Neill too was about to acquire a dog of his own.

One winter day Doug Ireland had a trip with the company's Anson to a point near Snowdrift on the southeast shore of Great Slave Lake. When he landed on Stark Lake the aircraft sank through the ice, and so Neill and Jack MacDougall immediately left from Yellowknife in the Stinson to see if they could retrieve it. With the help of one of the passengers on board they got the Anson up on top of the ice. "The props were pretty bent up," Neill says, "but when we got it running everything smoothed out and it appeared to be airworthy."

"Who's going to fly back with me in the Anson?" Doug asked. Neill volunteered. Doug had a little golden brown dog on board, part Cocker Spaniel and part Spitz, that he called Snoose. When the dog became restless during the trip back to Yellowknife, Neill picked him up and set him on his lap. When they got back to Yellowknife, the dog followed Neill everywhere. Doug lived some distance away, but that night the little dog made his way through the snow to appear at Neill's doorstep. Finally Doug said, "The dog is yours," and from that time on, Snoose accompanied Neill on his trips and became known all over the country.

Being an aircraft engineer in the far north was not an easy job. Herman Nelson heaters were used to warm the engines and get them started at the base. These military-made portable furnaces on wheels were equipped with a gas engine blower – a large five-pot burner operating on diesel or kerosene connected to 12-inch-diameter flex ducting. Smaller and simpler blow pots went with the aircraft for use in remote locations.

One time up at Coppermine with an Anson, a terrific wind and sub-zero temperatures made any kind of work miserable. The engines had been covered with canvas tents that came right down to the ice; a propeller blade pointing up served as a pole to create more height for working inside the tent. First thing in the morning Neill lit the blow pots, which threw quite a bit of heat, and placed one under each engine. Then he dived back and forth to check and make sure everything was going okay.

"When the heat started, the lower prop blade would ice up from condensation," Neill says. "By the time that frost had dissipated up to the engine hub, the engine would start. I couldn't start both engines at once so I'd take the engine cover off one, run around into the cockpit and start that engine. Then I'd leave it idling, come back down, and pull the engine cover off the second one."

But on this cold, windy, 40-below-zero day, the engines had cooled to the point where they just wouldn't start. Neill looked up to see the pilot on his way down to the aircraft, along with a priest and a couple of other passengers.

"I had sort of got mad by this time, with the cold and the engines not starting," he says, "so I took the two blow pots, which were still going because I thought I might need them again, and pointed them right at the engines so they blew flame right in them. I left that going for a couple of minutes or so, put the blow pots down, ran around and jumped in. Both engines were started by the time the passengers arrived.

The pilot, however, was less than impressed. "What in hell were you doing," he demanded. "Trying to burn that thing up?"

"There's no damage," Neill replied, "no burnt wires or anything. But don't think I wasn't mad enough to really consider burning it up!"

Neill and Jackie's third child, a daughter, Maureen, was born on October 25, 1951, in Yellowknife. That same year Associated Airways bought out Yellowknife Airways, and Neill just transferred over. "Sometimes the cheques weren't good for a few days," he recalls. "I guess they were a little tight for money after they bought out Yellowknife Airways. But the money always did come through."

Neill's independent spirit, gained from his years on the trapline, stayed with him on his new job. "One day I had some time off, so two other couples and Jackie and I were all ready to go to Taltheni Narrows, half-way between Yellowknife and Fort Reliance, to do some fishing. It had earlier been approved by management, but when Vern Simmonds, the operations manager for Associated Airways, heard I was taking off for a week or 10 days, he came down and said, 'You realize that maybe by the time you come back you won't have a job, eh?' I said, 'Vern, as far as I'm concerned I don't give a damn.' He stood there for a minute, and then he said, 'Have a good fishing trip. Your job will be here when you get back.'

"So sometimes you just have to stand up for yourself."

The Max Ward Wilderness Experience

Max Ward and a partner, George Pigeon, had started Yellowknife Airways in 1948, but Max sold his interest a year later and left the aviation field for a while. However, after Associated bought the company, Ward did come back for a short time to fly for them.

On November 19, 1951, Max was the pilot on a trip in the Bellanca Skyrocket from Yellowknife to Bathurst Inlet and Coppermine. Neill was the accompanying engineer, and their lone passenger was a Hudson's Bay Company inspector named Learmouth.

Their first stop was at Matt Murphy's camp, just a short jag off the direct route to Bathurst from Yellowknife. They visited for a while, picked up a half-dozen wolf skins, and left later than they'd planned in the afternoon just as darkness was beginning to fall.*

As they neared Beechey Lake they encountered a strong headwind, which convinced Max they would not have enough fuel to make it to Bathurst Inlet. "I think personally we might have been able to make it into Bathurst," Neill says, "but Max didn't think so, and as the engineer I had nothing to say about it. So anyway we decided to land near the Burnside River. We spotted the outline of what we thought was a lake, but the ground drift of snow kicked from this high wind had made drifts that were 20-25 feet high so we couldn't actually see the ground.

"From what we could see, the lake appeared to be round but instead it had a backward 'L' shape. There was a bit of a cliff running along and Max was using that as his guide. The Bellanca is a little bit 'blind' when you come down for a landing, and all of a sudden I saw this hill in front of us. I yelled to Max, 'Pull it up!' He did, but we were too far down at the time. We hit this hill and it kicked us up into the air. Max quickly shut off the motor and I flicked the switches off and stuff. If we had been over 200 or 300 yards we would have been all right, but instead ... we crash landed."

The airplane settled but the wheels had come apart at the cowling, and the wings and prop were bent out of shape. The Bellanca landed with one wingtip pointing skyward and the other touching the ground, with the disconnected wheels scattered out on the snow.

There was no fear of fire, so they carefully stepped out the lower side of the aircraft and onto the ground to survey the situation. A Mayday call to report their accident location was the first priority.

"We had to dismantle part of the antenna for the radio," Neill says. "Max loaded the radio – it was one of the Lears where you had to load the antenna on – so I was on one end of the antenna holding it out across the ground and he was in the aircraft. He gave his Maydays a few times. We weren't lost, we knew right where we were, but the company radio in Yellowknife was very poor with a terrific amount of static, so the base manager usually turned it right down to stop the noise. There was no answer from Yellowknife but Max called a few more times on the Mayday anyway, and gave our location. Yellowknife never heard us but Whitehorse picked it up."

The second step was to fix a shelter for what could possibly be a long stay. Here, Neill's survival skills kicked in. "On all those trips we made, I made sure there was a tent, extra socks, moccasins, mitts, stuff like that, and a big packsack full of food. Actually we had lots of food, and booze, as we were taking the Christmas supplies up to Bathurst Inlet and Coppermine.

---

* Max Ward also offers a graphic description of this trip in his autobiography, *The Max Ward Story*, pp. 94-101.

"We strung up the tent outside along the upper wing, using guy-wires to hold it up, and secured it along the ground with mailbags and frozen turkeys and hams. Then I opened a bottle of rum – but that was only after we got the tent set up."

The six wolf skins they'd picked up at Musk-Ox Lake were spread beneath their sleeping bags for extra warmth, and blow pots were set up for heating and cooking in the tent. Max told Neill later that one of his most colourful memories was of Neill mixing up cooked white rice with hands stained black with soot from the blow pots.

As soon as it got light the next morning they drained oil from the engine for fuel and a signal fire, and stamped large Xs in the snow. Neill worked to keep the conversation going over the five days they were stranded, essential to ward off depression and anxiety, especially on the part of their unfortunate passenger. Unsure when or if rescue might come, Learmouth took photographs of the scene.

Max Ward records his feelings about the accident with brutal honesty:

> *... I walked away from the aircraft, up into the rocks, where I could be by myself and take stock. I was 30 yeas old; I had a wife and two children, a pile of debts, and I had messed up my business, this flight, and very nearly three lives. I couldn't believe I had broken up that wonderful flying machine. I'd let myself make a beginner's mistake – I'd run out of time.*

An RCAF Dakota came over the day after the crash, and dropped a little parachute to ask if there was anything they needed. When the stranded airmen signalled that they didn't need anything, they left. (Neill later donated the little parachute to the Alberta Aviation Museum in Edmonton. It is displayed with the Max Ward memorabilia, which shows better times that followed, when Max went on to found the fabulous Wardair airline.)

When Whitehorse radio contacted Associated Airways in Yellowknife, the company sent up Abe Dyke and a prospector from around Yellowknife (supposedly for navigation purposes, although Max reports that he got lost en route) in a de Havilland Beaver. The cargo destined for Bathurst Inlet was loaded in the Beaver, along with Max Ward, Mr. Learmouth and the prospector, while Neill stayed with the crashed airplane. When Abe returned they loaded the Coppermine cargo, flew there, dropped it off, and then came back to Bathurst to pick up Max and the prospector, and back to Yellowknife.

"Anyway, we finished the job, just a few days late!" Neill says.

Neill flew back to the crash site the next summer to assess the needed repairs for the Bellanca, and Vern Simmonds and Jim Dick went to fix what was necessary to fly it back. "When they took off they just about wrecked it, because they hadn't realized that the low wing had gathered rain," Neill says.

"It had run down and become so heavy on the wingtip that the aircraft just about went down again."

Edmonton

Neill's aptitude for aircraft engineering had definitely sent him on a new career path. Then Jock Currie, a DOT inspector, came to Yellowknife and suggested to Neill that it was time he acquired some hangar experience. "I think he arranged with Associated that I was to go to Edmonton," Neill says, "so we moved and the company sent another engineer up to take my place. I had his house in Edmonton, and he had our place in Yellowknife."

Neill first worked at Aero Engineering, a subsidiary of Associated that performed major overhauls of engines, propellers and airframes. There he met Chuck MacLaren, who was servicing their bush planes. Neill still admires Chuck's method of assigning work. "I can't remember any time that he told me that I *had* to do something. He'd say, 'This aircraft needs another carburetor,' and that's all he had to say. I'd go over and put a new carburetor on it."

"At one time a basic engineering endorsement would let you work on pretty well anything up to 6,000 pounds, and then you had to get an endorsement for extra ones," Neill states. "The first one I got for larger aircraft was on the old Barkley-Grow, a twin-engine all-metal eight passenger monoplane. Only 11 were ever built, and Associated had three of them. I carried on from there to qualify for different endorsements and get my licence."

The time in Edmonton proved to be bittersweet. He gained his essential engineering licences, but his marriage broke up. With his priorities suddenly – and radically – changed, he decided to give his notice. Then, Vern Simmonds confronted him and said, "What's this about you quitting your job?"

"So I told him why," Neill said, " and he asked where I was going. I said I thought I'd go back to Peace River and do something or other, because I had the three kids to look after. He said they had some work out of there, and a base, and asked how I would like to go up as engineer. I said fine, so they flew me and the three kids up with all our stuff, and I was back home."

Peace River

Associated Airways had a contract with Gulf Oil to provide a Beaver aircraft and a pilot for work out of Fort Nelson. The Beaver, originally owned by Imperial Oil, had the unique registration of CF- OIL. When Imperial wanted the registration back, Neill spray-painted over the old letters and taped on the new ones, CF-GQN.

The pilot was to return periodically to the base in Peace River to have the checks done, but not long after Neill arrived in Peace River the pilot "got himself into a bit of a problem" with Gulf Oil. Vern Simmonds asked Neill to go to Fort Nelson and find out what seemed to be wrong with the aircraft – which, according to a report from the pilot, had just sunk in the river – and to stay there for the summer to monitor the contract. Then Vern added, "You tell that pilot exactly what he has to do to correct the situation, and keep good relations with Gulf."

"That's the first time I ever heard of an engineer telling a pilot what to do!" was Neill's response.

"If he objects, you let me know," Vern said sternly.

Neill made arrangements for Terry's and Peter's care, and baby Maureen stayed with Neill's mother, while Neill flew to Fort Nelson.

Dr. Andy Bailey, the geologist from Calgary in charge of the exploration project, arrived in Fort Nelson on the same CPAL flight as Neill. As they drove down to the river in a pickup truck, he let Neill know that he was "red hot" about the situation.

There was the sad looking Beaver, sitting on the river across from Old Fort Nelson, with its floats completely submerged. "It was tied up there, and when even a canoe came along it would slop water onto it," Neill says. "So water got in the floats and they sunk."

The first job was for everyone to take turns pumping out the floats to see if there were holes in them as the pilot had reported. "It's quite a job pumping out two floats that are completely full of water with a doggone little pump," Neill says. "But about 10 guys were there so we got it floating and nothing seemed wrong with the floats at all. I said, 'Okay now, everybody jump on the floats, get some weight on them.' We stood there for 15 minutes and couldn't get another drop of water out of them. 'As far as I'm concerned there are no holes in those floats,' I said. 'The only thing wrong with this aircraft is the fact that the pilot never pumped the floats out.'"

Andy Bailey turned to the pilot. "So, what's the word?"

The pilot shrugged.

"By the time you get a load ready to go to Nahanni Butte, I'll have had a quick look at the aircraft," Neill said.

But the pilot wasn't ready to lose rank to an engineer. "Now just a minute – that's up to me," he said.

"No, it's up to *me*!" Neill responded. "You can check with Vern Simmonds if you don't think so."

The aircraft was loaded and successfully made two trips that day.

Nahanni

Neill found the Nahanni region to be beautiful "if you like that kind of country." Being more used to the tundra, he found that mountain flying carried certain risks. Two of the company's head office men from Calgary were on board one day and wanted to see some of the rock formation up close. The pilot, likely under stress from his earlier chewing out, obliged.

They were flying along when suddenly the aircraft dropped down. "What the hell are you trying to do?" Neill asked in a low voice, trying to not let his anxiety be overheard by the brass sitting behind him.

"They wanted to see the formation!" the pilot replied.

"Not at this altitude! You'll have a hell of a time getting out of here! There are cliffs on both sides! I don't know what the downdrafts are or anything else here."

The pilot reconsidered and looked for an opening. But by this time the company men had become aware of the danger they were in, and endured a very tense and silent flight through the canyon before the pilot found a spot where he could get out.

"That didn't endear him too much to the brass," Neill concludes. When further problems arose, the pilot was let go and a replacement pilot did the rest of the flying on the Gulf contract that summer.

Neill was informed that the project manager was no longer willing to sign a contract unless Neill Murphy was the engineer. "That gave me a bit of an ego boost – but it didn't do a damn thing for my salary!" Neill laughs.

Nahanni Butte was Gulf's first base camp out of Fort Nelson. Dick Turner, a local resident, had a place downriver and at times Gulf would charter his boat, and hire him to do camp moves on the river.

The peaceful lives that these residents had enjoyed in the remote but famed "Headless Valley" of the Nahanni, similar to Neill's earlier existence in the Barren Lands, were soon to be drastically changed as the area was inundated with mining and mineral and oil and gas companies seeking to exploit its wealth.

Nevertheless, it was still possible to enjoy moments of relaxation. "The first year we were at Nahanni Butte," Neill recalls, "one of the pilots brought us a couple of bottles of over proof rum from Yellowknife, one for me, one for Charlie Ferber, the helicopter pilot at the time. We each had a drink in the evening after all the flying was over.

"The next day Ferber went flying and I went on a trip in the Beaver. When we came back that evening, Ferber looked at his bottle and said, 'Murf, are you drinking my rum?' I said, 'No, I've got my own.' So he drew a mark on his bottle, and the following evening, when he lifted the rum bottle to check the level, it was up to the mark all right, but there were bugs floating around inside the glass!

"What had happened, the cook had taken a nip out of it and filled it with some water from the creek, and hadn't strained it. So we knew the cook was having a nip. We hid our rum after that."

Fixed and Rotary Wings

In the early 1950s, Associated Airways realized that helicopters would soon become essential machines for exploration and surveying work in the oil industry. They formed the wholly owned subsidiary Associated Helicopters Limited and purchased a Bell 47D1. This model seated three persons (one seat held the two passengers side by side), and featured a clear one-piece acrylic plastic canopy bubble and an open tube-work tail boom. The company asked Neill if he was interested in coming back to Edmonton to add to his Aircraft Maintenance Engineer endorsements for rotary wing aircraft.

"At that time there were only two outfits that operated helicopters," Neill says. "One was Okanagan Helicopters and the other was Associated Helicopters, who by this time had two machines. They had wrecked one at Trutch, Mile 178 on the Alaska Highway, so my training was to help to rebuild this helicopter. They brought the wreck in and I learned how to repair it, and got my endorsement."

One incident from that summer has stayed crystal-clear in Neill's mind. "We were camped at the mouth of the Mountain River where it runs into the Mackenzie. The regular pilot for the Beaver had left for a couple of weeks and they got another fellow in, Earl Lozo, to do the flying with the Beaver. The geologist in charge, Dr. Andy Bailey, who was at the camp at the time, was a bit nervous about the new pilot and asked me what I thought. I said, 'I know him, he was flying helicopters, but I don't know how he is on fixed wing.' So Andy asked if I'd mind going with him to take three fellows out to one of Gulf's fly-camps."

Like many aircraft, the Beaver was equipped with an inertia starter. Instead of a direct drive starter similar to that of a car, the inertia starter had a smaller electric motor that would wind the flywheel up to 15,000 rpm. The flywheel momentum, when transmitted through reduction gears to the engine crankshaft, would turn the engine over for perhaps two revolutions. If the engine didn't fire or start in these two revolutions (which frequently happened) then the whole sequence of priming and winding up the inertia flywheel, which would take 10 to 20 seconds, would have to be repeated.

Neill explains the set-up with this particular Beaver aircraft. "There were two switches on the instrument panel and another toggle down below with piano wire running down to the starter. You flicked one of these switches and this inertia starter wound up like a cream separator – *r-r-r-r* - until it got going. Then you flicked the other switch and it kicked in a solenoid that booted the starter. And then of course the engine would start."

"We had told the new pilot time and time again to never push the aircraft out into the river until the engine started. But he gave it a shove and out we went. The Mountain River is very swift and the Mackenzie was going at pretty good steam at that point, too. The current caught us right away, and it was only a matter of a few yards until we were into the Mackenzie.

"As I said, Lozo wasn't that familiar with the Beaver. He flicked the inertia switch, and then the second one, but nothing happened. The kicked-in solenoid didn't boost the starter. I said, 'Wind it up again,' which he did. When it was whining away, I said, 'Okay, there's a lever down there, just grab that toggle and pull it.' Well he was so excited by this time that he gave it a yank and it pulled the doggone wire right out of the starter, so we couldn't start the thing. By this time we were going down the Mackenzie River, on floats. Fast. To top it all off, the Sansu Rapids were only about a half-mile away."

Neill turned to the three passengers in the back. As calmly as possible, he told them, as a precaution, to take off their heavy climbing boots, make sure they each had one of the self-inflated life jackets that the Beaver carried, and to open the door and latch it so it stayed open. He then called to the pilot "Switches off!" and jumped down onto the float. He cranked the prop with one hand, and with the other hung onto the ski fitting on the side of the airplane, which wasn't very big. When he got the prop turned to build compression, he yelled, "Switches on! Throttle set!"

It didn't work. He tried it again. Didn't work.

"By this time the rapids were ready to chew the ass off this Beaver," Neill says. He had time to try to turn the prop once more, and then he'd have to yell at everyone to be ready to get away from the airplane. He gave it another good swing. It caught.

"I don't know how much Lozo had on the throttle, but when that thing started it was just one cloud of water! I was ahead of the strut and I just hung on or I'd have been blown away into the water. Of course I got soaking wet."

When the pilot got the aircraft turned around, he throttled back a bit so Neill could jump into the front co-pilot's seat. The doors were closed, so they just taxied back up into the Mountain River to dock at the camp. Neill fixed the starter and they left again to take the passengers out to the fly camp.

"I tell you, it was a bit of a trying moment, if you want to call it that," Neill says. "And Earl Lozo learned a little bit about a Beaver in a hurry."

Dick Turner later said he'd seen the Beaver hurtling down the Mackenzie, and had started his boat to try and catch them. "But there was no way he could have, before we hit the rapids," Neill says.

Earl Lozo unfortunately suffered a fatal accident some time later, when another company brought him in for two weeks to replace their regular pilot on a Beaver.

The company Beaver (GQN) also met with disaster in the Arctic when the pilot missed his takeoff, hit a wave and the airplane sunk. "It was eventually salvaged," Neill says, "but by that time the salt water had chewed holes through the engine, so it had its day."

Blue Skies

Neill's personal life took a positive turn in 1952, and all because of bad weather and a health issue. He had been working in Edmonton for a few days, but when he was ready to go home CPAL had cancelled their flights to Peace River because of the weather, so Neill took the train. A nurse, Willemine (Pim) Pruyser, was also on the train on her way back home to Peace River. They exchanged friendly greetings, and she told Neill that a few days earlier she'd flown with a patient to Edmonton on a medevac mission.

Pim had received her nursing training in her home country of Holland. Being an adventurous type of person, she'd moved to Sweden for a couple of years to learn the language and practise nursing there. When her brother moved to Canada and settled in the Peace River area she decided that might be her next destination. She was now working at the Peace River Hospital.

A few days after this auspicious meeting, Neill developed appendicitis and found himself not only in the same hospital but under Pim's care. In these odd but serendipitous circumstances, their relationship developed. When Neill was asked by Associated Helicopters to move to Edmonton, he rented a house there and prepared to move his family under one roof. And then he proposed marriage.

Neill and Pim were married on November 27, 1954. Their daughter Theresa was born July 21, 1957.

Eldorado

Three years later, in 1960, Neill left Associated Helicopters to work for Eldorado Aviation. The company had bought a Sikorsky S-55 for the mine at Uranium City in northwest Saskatchewan and for use during construction of hydro lines at Wellington and Waterloo lakes. Neill was offered the engineer's job, and was sent to take a course with Pratt & Whitney in Montreal to get his endorsement. *

* Atypically for a helicopter, the Sikorsky S-55 was equipped with a modified version of the Pratt & Whitney R1340 radial, basically the same engine as used on fixed-wing aircraft such as the Norseman.

Then he returned to Edmonton, and made preparations to set himself up at the base in Uranium City. "I first had to go to Winnipeg, put two overhauled engines in each helicopter for them, and do maintenance checks on both of them. We did a little bit of work around northern Manitoba at Norway House and in through there."

The eastern-based company that owned the two helicopters contracted to Eldorado had been working on the Arctic Shelf, and also held a contract for a couple of helicopters to work on the DEW [Distant Early Warning] Line. When they relinquished the latter contract Eldorado acquired it, and arranged for Neill to be the engineer on that project as well. The plan was to send him to work on the DEW Line for the summer, and then back to the base at Uranium City. When he finished the initial work in Manitoba he made a brief visit back to Edmonton and then left with the helicopter crews for the DEW Line.

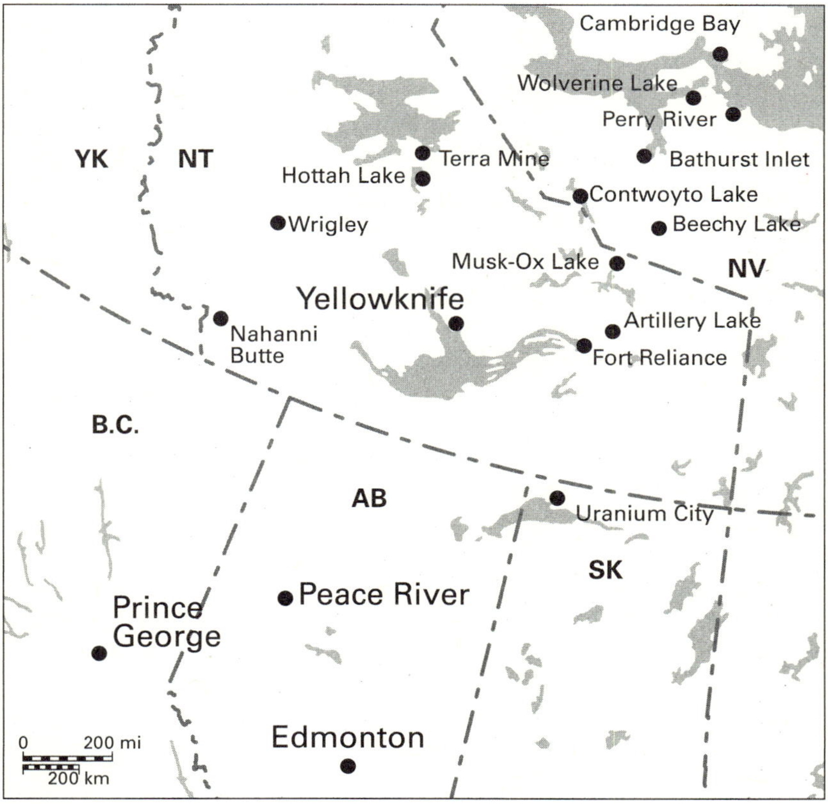

The World of Neill Murphy

"They had a helicopter on each of the two U.S. Navy ships, called LSTs [Landing Ship Tanks; these were small vessels developed in the Second World War to land tanks and trucks on beachheads], that were supplying the DEW Line. Each ship was outfitted with a tank on the bottom, and an area for freight on the middle deck. The helicopters and trucks or anything else they were hauling were parked on the top deck. Clamshell doors opened to a ramp that ran right up to the top deck. The company employed a couple of pilots, an American and an Englishman, to fly the helicopters and I spent that summer running back and forth from one ship to the other. They had no covers for the helicopters so I smeared grease on them anywhere I thought they could become corroded from the salt air."

The two ships travelled along the coast, docking at various sites to open their clamshell doors, and drop the ramp. Some union men who were working on contracts for other companies were to load or offload the freight or equipment. Neill was not impressed with the work ethic he perceived. "When we pulled in to one of these places where we'd have things to unload for the DEW Line site, these men would come out, take a look around, and then go back in and hit the sack aboard the ship. The native fellows who were working for the DEW Line did all the work with the Cats, smoothing the gravel and everything up to the ramp, and these other guys were on board the ship, taking it easy. The only time you'd see them was when there was food on the table. So that's the way they treated the locals."

This and other situations with the company did not sit well with Neill, and his time with Eldorado was not particularly satisfying. There were of course some "good guys" such as Chuck MacLaren working for the company, but with the corporate heads all related in one way or another, and Neill being "odd man out" since he worked solely on helicopters while the others were "fixed wing fellows," he decided that four years with Eldorado Aviation were enough. He resigned in 1964.

Northward Aviation and Peace Air

The family moved back to Peace River, where Neill quickly found work with Northward Aviation from 1965-67, and then for Peace Air from 1967-69, on fixed wing.

He produces a photo of an interesting-looking upside-down aircraft, a Northward Aviation Beaver (CF-ITE), and explains, "This airplane was chartered to Forestry. The pilot, a fellow by the name of Bernard, figured because he had big wheels on it he could land anywhere, so he tried landing on the airstrip at Worsley, which was a Forestry base at that time. But there was three and a half feet of snow and that's the result. He came in and flipped right over. It broke the fitting on the right wing."

They righted the aircraft onto its wheels, inspected it, removed the wing and sent it to Edmonton by truck. Three weeks later the wing came back. They installed it, along with a new propeller, and one of the pilots flew it to Edmonton.

"They put it in the hangar and that night the hangar burnt down along with the aircraft," Neill says. And that was the end of CF-ITE.

Many pilots and engineers go through frequent changes from one company to another as opportunities rise and fall – a chance comes to work on new and different aircraft and so they leave to improve their qualifications; a company can't offer enough work to keep an employee year-round and he has to go elsewhere; or sometimes the personnel or safety issues at a particular place become intolerable. Aviation companies can roller-coaster through an amazing succession of financial ups and downs, especially when many small charter airlines start with borrowed capital and go on "a wing and a prayer" as they try to keep one or two aircraft aloft and meet a payroll.

Neill Murphy's career in aviation was no exception, as he went from Yellowknife Airways to Associated Airways and Associated Helicopters, to Eldorado, to Northward, to Peace Air and finally to Gateway. There the real excitement began, for Gateway was in bad shape, producing for Neill many challenges, and not a few headaches.

Gateway Aviation

"I was still working for Peace Air, but one day I was in Edmonton and decided to go to the Gateway Aviation hangar to visit a couple of fellows there. They said I should go and see the chief engineer, a fellow named Sokolosky."

When Neill entered his office the man was on the phone. Suddenly he looked up and said, "There's one just walked in the door!" He immediately hung up and said to Neill, "We need an engineer. I want you to go to Yellowknife."

Neill could only respond, "But, I've got a job."

But Sokolosky made him an offer he couldn't refuse: the position of base engineer at a much higher wage than he was making at Peace Air, a free family move, and a company house in Yellowknife.

But a shock awaited Neill when he arrived at Gateway's base in Yellowknife: six airplanes were tied up to the dock – a couple of Beavers, a Single Otter, a Cessna 180 and a 185, and a 206 – all unserviceable.

Bob Warnock was the base manager, and also one of the pilots (although not legally – he had lost his licence because he had leukemia, but he was still flying).

"What in hell is going on?" Neill asked as he walked along the dock past all the inert aircraft. Warnock explained that Gateway had hired a company to do their aircraft maintenance on a retainer basis, but they'd do a bit of work and not complete the jobs. Neill was stunned. "The only thing I can do is to ground them all," he said, "and then I'll work on them one at a time."

"They can still fly!" Warnock protested.

Neill was adamant. "No. They're going to stay here." He then grounded the entire fleet.

As he began working on the aircraft, a second problem emerged: lack of business. The airplanes had been sitting at the dock and everybody in town knew they were unserviceable. There was one active contract for the 206 with one of the fishing lodges, but that was all.

"We had three, four, and sometimes five airplanes still tied to the dock, even though they were serviceable by that time, because of the company's reputation," Neill recalls. "Bob Warnock stayed around for a while and when he left he gave me the job of base manager. He finally left Yellowknife and died.

"So I went around to see everybody I could think of and assure them that everything was fine, and finally business started to pick up a little bit. The next summer we could hardly keep up with the business. At times, among all the aircraft they would do 100 hours of flying in a day."

When the work for the two Beavers fell off, the company took them back to Edmonton. But when Neill kept receiving customer inquiries asking for the Beaver he contacted Edmonton and convinced them to bring one of them back to Yellowknife. To his delight, he received Serial #1 (registration CF-FHB), overhauled, newly painted and looking great.

"We had a regular run to Terra Mine at the south end of Great Bear Lake, a real good trip because it took a couple of hours to get up there and a couple of hours back. Anyway, I had two trips this one day for our Otters, and also a trip for an Otter for about 30 miles from Yellowknife to a fish camp. A group of Americans had come up and were waiting to go out in the Otter. I was in a sweat about how I was going to accommodate them with the Otters being fully booked, so I asked them if they'd ever seen the first Beaver that was ever built.' No, they hadn't. 'Well, there it is, come and have a look.'"

Neil explained that it looked new because of the paint job; then he opened the door to reveal the stainless steel plate with the registration, serial number and year it was built. They were duly impressed. Then Neill made his pitch. "If you fellows will let me have that Otter for a trip we need it for, we'll split your group and send you over in two trips with the Beaver. Then you can say you flew in the first Beaver that was ever built."

They readily agreed. He loaded the Otters, they went off to Terra Mines, and the Beaver made the two short trips with very happy customers. "I used FHB on numerous jobs," Neill said. "It became an old friend of mine."

Finally, however, Neill was forced to give up on Gateway Aviation. "They had some accidents, one with the Beaver east of Yellowknife, and then that Marten Hartwell business …"

The Hartwell Horror

In 1972, Neill developed a heart fibrillation and for a while would go to work only in the afternoons. "A trip came up for Cambridge Bay on November 7, a day when the weather wasn't very good," he says. "Marten Hartwell, one of our Gateway Aviation pilots, was supposed to take it in the Beech 18 [CF-RLD]. I told him to make sure he had certain conditions at Contwoyto Lake before leaving, and to check the weather. He was sort of a belligerent bugger, he didn't talk too much, but there was also that old deal, you know, where the engineer is not going to tell the pilot what to do. 'We'll do what we want' – so that's what he did."

Neill had voiced strong objections to the company hiring Marten Hartwell. He was, after all, the base manager for Gateway when Hartwell was hired and felt he should have a say in it. "I hollered my head off about it, but Doug Rae, our Operations Manager, said, 'Oh no, he's fine.' He had so many hours from the German air force, the Luftwaffe, whether it was on Twins or what, I don't know, and he'd flown an Otter for us for part of the previous summer for a geodetic survey. They complained about him, saying he was ordering the students, and even the guys running the project, to roll out the gas barrels and gas his airplane, stuff like that. I offered to give them another pilot, but as it was near the end of the season and they were almost finished the job they wanted to leave it as it was."

When that job was over Marten Hartwell went to Edmonton to see the Gateway managers. One of their pilots had just quit to fly for Wardair and now Gateway had a Beech 18 without a pilot. Marten got the job.

"When I heard he was coming up to Yellowknife to work for us, I said, 'My God, what's the matter with your head?' 'It's perfectly all right,' Doug Rae said, 'We checked him out.'" Neill shudders at the memory. "They checked him out all right."

So Marten Hartwell left in the Gateway Aviation Beech 18 on November 7, 1972, and the rest is history.

Marten Hartwell [a.k.a. Leopold Herrmann] had received his flight training with the military in Germany in 1944. He augmented that training to gain a West German private pilot's licence in 1958, a commercial pilot's licence in 1967, and multi-engine and night-flying endorsements in 1970. However, he was still restricted to flying VFR [Visual Flight Rules], which meant that he was not qualified to fly with passengers at night when he couldn't see the ground. By the time he left out of Yellowknife for

Cambridge Bay he had accumulated 2,085 hours in the air, 25 of which were as captain on the Beech 18.

Neill picks up the narrative. "He left and got himself lost on the way to Cambridge Bay, went a way out over the ocean – and this was in the wintertime. Luckily he was with Gren Thomas, a promoter and prospector. He finally got Hartwell turned around and they got back to Cambridge Bay."

Ed Logozar, a senior pilot with Gateway Aviation who was flying their Twin Otter, had come in along the coast via Spence Bay (Taloyoak) on the Boothia Peninsula. There he'd picked up a medevac destined for Yellowknife: an eight-month-pregnant Inuk woman, Neemee Nulliayok; a 14-year-old Inuk boy, David Pisurayak Kootook, who was thought to be suffering from appendicitis; and a government nurse, Judy Hill. Ed Logozar brought the passengers as far as Cambridge Bay, but as he was then scheduled to go to Coppermine it was decided that Hartwell would take them on to Yellowknife. It would be dark before he got there, and Hartwell was qualified to fly only VFR, but these dangers were apparently not considered.

"If Hartwell hadn't been there, Logozar probably would have brought them right to Yellowknife and they would have been fine," Neill says. "Or, if I had heard about this, I probably would have got Logozar to bring the Beech 18 down, because the weather wasn't any good, and then fly those people back with one of the other pilots who was in Cambridge Bay. Instead, he turned everything over to Marten Hartwell, and Hartwell left with these passengers for Yellowknife."

What happened next became one of the saddest aviation tragedies in Canadian history. *

Neill has his own ideas about how things went so tragically wrong. "Hartwell never mentioned it except very vaguely, but I think what happened is, he had his radio on and he locked onto the signal from Wrigley instead of Contwoyto Lake. There's only one dot or one dash difference – I'm not sure which – on the Morse code signal."

Wrigley, located on the east bank of the Mackenzie River just below its confluence with the Wrigley River, is about 250 miles northwest of Yellowknife. The distance between Cambridge Bay and Yellowknife is nearly 470 miles. The signal at Contwoyto, which is halfway between Cambridge and Yellowknife, is the one Hartwell should have chosen for that part of the journey.

"I think he was locked onto the Wrigley beacon," Neill continues, "because when he hit this hill near Hottah Lake he was on the east side of Great Bear Lake, which is almost on a direct line from Cambridge Bay to Wrigley.

---

* The event was chronicled in numerous newspaper accounts, and also in a book *The Survivor: The Marten Hartwell Story* by Peter Tadman.

"I think what might have happened, my idea anyway, is that when he realized he shouldn't be seeing Hottah Lake, maybe he turned the lights on in the cockpit to look at the map; maybe he got the nurse to hang onto the stick or whatever; or maybe he pushed the stick forward with the map on his lap. Anyway, he hit a hill, a knob that was sticking up. It killed the nurse right away, and the other woman in the back. The young boy survived, and Hartwell, but his legs were broken."

The crash site wasn't discovered for 31 days, while the temperature hovered at -35° F. David was able to set up a tent and crawl around to gather twigs for a fire while Hartwell, immobile because of his broken legs, could only give instructions.

"The boy couldn't talk English," Neill says, "but maybe Hartwell got him to get wood and stuff by yelling at him even though he couldn't understand. And the little fella was really out of his element because they were in fairly heavy timber and the kid had never seen a tree before in his life, and he must have been pretty scared."

Both David and Marten described their dilemma on notepaper, and these letters became instrumental later to help others comprehend the terrible, and for David, fatal, circumstances. When their food ran out and starvation seemed the only option, Hartwell made an incredibly difficult decision: to survive by consuming the flesh of the dead nurse. David refused.

"I think he got that fatalistic attitude, 'there is no use of me being here,'" Neill surmises, "and so he died."

The wreckage was finally found on December 9, 180 miles off course and 250 miles northwest of Yellowknife. All were dead but the pilot, Marten Hartwell.

He solemnly described the ordeal for the media, and emphasized the bravery of the young boy in helping to keep him alive. (In 1998 David's memory was honoured by the Northern Transportation Company by having one of its ships named after him.)

Because of his position with the company, Neill was subpoenaed to testify at the court inquest and asked for his opinion on what might have caused the accident.

"I told them what I'd mentioned about him not going in this weather, and I think that was brought up," he recalls. "The judge also asked me about the survivor kits the aircraft carried. I said that our aircraft were equipped when flying in the North with two survivor kits that each had enough food to last two people for 16 days."

Neill also informed the court that Hartwell was familiar with Hottah Lake, just 12 miles from the crash site. In fact, Hartwell "had landed on the lake three months previously and had flown extensively in the area in the months leading up to the fatal mercy flight." Neill said he had also given

Hartwell specific instructions the day before the crash on how to react in adverse weather conditions.

When questioned about the icing, low weather conditions and poor visibility that Hartwell reported encountering that day, Neill testified, "My instructions were to him to the effect not to attempt another trip until he had at least a 2,500-foot ceiling and 10-mile visibility, and I explained why."

He seemed to understand the instructions, Neill said, but when asked if he had done what he was supposed to, Neill commented, "Obviously not."

That said it all.

End of the Line

Neill's departure from Gateway Aviation came about by mutual consent. "I said good-bye to them, and I think Gateway said good-bye to me too. They claimed that there hadn't been an entry made in a logbook in regard to a check – a 25-hour inspection on a Twin Otter. I hadn't checked it off because there were another couple engineers there so I didn't look at the book. I figured the guy had an engineer's licence, he should have done that."

Pim and the kids had come to live in Yellowknife while Neill worked for Gateway Aviation until 1976, and stayed there during his next job with Aero Arctic Helicopters, still based in Yellowknife. He was that company's engineer from 1976 to 1978, servicing their Sikorsky S-55 turbine and two Bell 206s.

His ultimate decision to leave Yellowknife meant giving up his Northwest Territories Wildlife Service hunting and trapping licences, which he'd kept all the time he was in the North, as well as his prospector's licence. It was the end of an era.

He and the family moved back to Peace River, where he worked for a year or so on a retainer plus an hourly rate for a Montreal company, Cardinal Helicopters. Following that, he worked for Peace Air on the same basis.

"In the early 1980s, I more or less gave up my licences and retired from aviation."

By that time, Neill's M-4 licence authorized him to work on everything from gliders to turboprop airplanes, and actually included many types that he'd never even seen before. He was qualified to work on all the smaller fixed wing aircraft, and then the Barkley-Grow, single and twin engine Otters, the Beaver, and the DC-3. When Gateway got a Mitsubishi MU-2, a high performance twin-engine turboprop machine, used for medevacs out of Yellowknife, Neill became qualified to service that as well. His engineering licences for rotary aircraft included the Bell 47, Bell 206, and Sikorsky 55 and Sikorsky 55 turbine.

"The qualifications and skills that make a good aircraft engineer are, basically, that he should like what he's doing," Neill says. "He has to have a good company to work for, with input from the employer and help from other guys. That gives him a feeling of wellbeing. Of course he has to have training to start with. If the training is not there, somebody has got to be with him and show him, or he's going to make mistakes and everybody suffers."

Neill Murphy has lived a life that few in the modern world could emulate, or perhaps would even want to. He's survived a tough existence in a tough climate, and learned from experience how common sense and some survival savvy can be one's saving grace.

If you become lost in the wilderness, he advises, "Don't move. If you're with an aircraft, stay with the aircraft. If you're with a snowmobile or vehicle or whatever else, stay with it, because that's what they're looking for. They're not looking for you running through the bush."

He refers to his old friend Johnny Bourassa, who experienced a forced landing in a Yellowknife Airways Bellanca in 1951 at Wholdaia Lake, NT, about 325 miles southeast of Yellowknife, and left a note in his downed aircraft saying he was going to walk out.* Neill was called in to assist with the search, and it is his lifelong regret that Bourassa was never found.

"Johnny stayed there five days, but what good did it do to leave? He should have stayed there all summer. He must have known that eventually somebody would run into him. The search was called off in the spring and the aircraft wasn't found until fall, but he could have survived. The ration kits in aircraft carried snare wire that could be used to catch rabbits. And there are also plants and berries. Most mushrooms are good to eat. If they look good, they're good to eat. Caribou eat them all the time. There was fishing gear in the aircraft and there were fish in the lake."

Neill has never encountered danger from wild animals in all his years spent in the wilderness. "The only time I got scared from a bear was a Barren Lands grizzly that I shot. You don't see too many of them, but they are there. It went down with all four legs spread out and his head lying on the ground. I waited a few minutes and saw no movement, so I went over to him, standing to one side. All of a sudden his head came up, mouth open, and he gave his god-awful roar. It didn't take me long to put another shell into him. And then I skinned him."

Old Huts and Haunts

In 1985 Neill revisited the old camp near Musk-Ox Lake, with his son Terry and daughter-in-law Lynne.

---

* Chronicled in "Where's Johnny?", pp. 122-151 in *Lost: True Stories of Canadian Aviation Tragedies* by Shirlee Smith Matheson.

"It was kind of sad, in a way," Neill says. He pauses a moment, recalling his year-round existence in the Barren Lands, and the complex balance required to survive and enjoy its bountiful beauty.

He produces a photo taken of the remains of the little caribou-hide hut. In the 60 years since his departure it has sunk gently into the permafrost, with just some poles, a few piles of sod, and a scattering of sun-whitened caribou antlers left to mark the spot on the tundra that he, his father Matt, his wife Jackie and son Terry had once called home.

"If circumstances were different … I'd go back there in a minute," he says, with a wistful smile.

The small lake just down the hill from the hut is little changed; the caribou and wolves still run along the brow of the far hill; the Inuit, the people born in this land, are less nomadic but still inhabit the area.

But Neill made a choice all those years ago, and built on his knowledge of survival as a Barren Lands trapper to add a new and exciting element to his life in the ever-changing field of aviation. Still, those memories are precious, and for Neill Murphy, and his father Matt before him, Musk-Ox Lake was a special time and place, and a way of life like no other on earth.

The 'Rev' Hal Tetz – Helicopter Pilot

"How I got into aviation was, I had been thinking about it all my life. Years ago there was a radio program called The Jimmy Allen Show, sponsored by BA Oil [British American Oil Company]. Jimmy Allen would pretend to be a pilot, flying into dangerous places. I'd listen to the adventures and hear all these airplane noises, and I got so I could hardly think about anything else but airplanes."

The early life of Hal Tetz indicated little promise that he'd ever follow his dream of flying. But as one adventure led to another, and he amassed skills in everything from blacksmithing to bartending, the dream that never faded would one day become a reality.

Hal was born in Bassano, Alberta, in 1927. When he was five years of age his family moved to a farm near Lacombe. "My dad was not a farmer," Hal says. "In Russia he'd worked with the railroad as a blacksmith, and so he started a little blacksmith shop on the farm. Finally the family moved to the Rosedale Valley two miles from Lacombe, and Dad started a shop in the town."

Early memories crowd in: being one of eight children in a "dirt poor" family, working from an early age at anything one could find "so they didn't have to feed you at home."

The family, being Seventh Day Adventists, sent the kids to the Canadian Union College up the hill. Hal attended their church-sponsored school from grades one to part of three, but the fees became a financial burden so

Hal and his brothers and sisters were sent to the public school in Lacombe, which was free.

His first paying job at age 13 was for a farm neighbour during fall threshing season. He'd start work at 6:30 am, harness four horses and bring them out to the field where the morning crew was waiting, and bring the alternate four-horse team back for feed and rest. The day would be filled with myriad chores. Then at 4:00 pm he'd brush, harness, and drive that team back out again and return with the first team, while the crew cut and stooked grain until dark – all for 50 cents a day. School was forgotten for the harvest season, which didn't make Hal that unhappy. "I went to school mainly because the governments made us stay until age 15," he says, "and then I was gone."

Seeking Adventure

One year, when harvest was finished for the season, Hal's brother-in-law, Webb Frizzell, was offered a job with the No. 32 British Commonwealth Air Training Plan [BCATP] Elementary Flying Training School at Bowden. He got the kid hired on, too. Hal was all of 15 years old.

"Bowden was the place where I got to know about airplanes," Hal says. He worked as a ramp rat, spending full days pushing airplanes in and out of the hangar, fuelling, and swinging props to start the engines. It was here that his early love of aviation became a reality. "I was nuts about airplanes, especially when I got to go up with the pilot. They had to do a 15-minute test flight after the aircraft were signed out by the mechanic and I'd get to go on these short rides. Every time they needed an L-14 signed (a technical log book indicating that certain checks had been done) I would grab a parachute and go along with the instructor. I was really hooked on flying, and when I was anywhere near an airplane I would help them to load and unload and so on. Pretty soon a lot of people got to know who I was and became friends.

"When I got to Bowden they were just getting rid of the Stearman airplanes, and had deHavilland Tiger Moths. When I left they were switching over to Fairchild Cornells."

After Bowden, Hal went to Edmonton to work at the BCATP No. 2 Air Observer School, again as a ramp rat, where he came to know the Mark 1 and Mark II Avro Ansons.

While in Edmonton he heard of a government organization called Selective Services that listed available jobs in Canada's north. The company would fly employees from Edmonton to the site as soon as they signed up. There was an opening for a cook's helper in Yellowknife, and so, in 1942, the 15-year-old packed his bags to work at the newly developing Giant Mines.

A series of hard labour jobs followed. After getting his driver's licence at age 16, he drove a truck, laying gravel for what is now the service road to the Yellowknife airport. "Three of us lived in a little tarpaper shack where the main town is now – there was no town there then, just the old town. In the winter, when I'd wake up in the morning, I could see all the nails that came through the boards – it was so cold inside the shack they had balls of frost on them. We took turns getting up every hour-and-a-half to start the three trucks so the gearboxes wouldn't freeze up. We'd start one, run it downtown and around the island, come back, park it, and get in the other one and do the same so we could work the next day. We also took turns getting up to light the fire in the morning. It was a tough job."

He then hired on with the Sheck Brothers to run a D-4 Cat, pulling two sleighs of cut wood. "There was no bulldozer blade on the little Cat and I kept getting into more and more trouble," he recalls. "I had to knock down little trees with the tracks to get the Cat twisted around and out of there. I finally got back into Yellowknife at 10:00 at night. It had been dark for hours, and I was frozen and hungry. I left the Cat running – had to or it wouldn't start in the morning – and I never went to work for that company again."

His next job was no easier. He worked underground "hand-tramming" – pulling ore out of chutes, filling the carts and then dumping it down another ore pass – for Consolidated Mining & Smelting.

Several other jobs added to his resume. While playing softball one summer at Moccasin Flats, the big Swedish bar manager who worked at the Yellowknife Hotel asked Hal if he could come over to the hotel and give him a hand. Knowing he was underage to work in the bar, the manager handed him a parka and sent him into the back cooler to stock shelves. At that time, Hal recalls, the miners drank beer in pints (35 cents) or quarts (70 cents). One day in walked two Mounties, who also played on the ball team. "Here I am," Hal thought, "a Seventh Day Adventist, under age, in a bar, and these two Mounties know me! I'm dead!" He did his best to stay out of sight as the policemen each picked up a pint, sat down on a couple of boxes, drank their beer ... and left.

But flying was still on his mind. When he heard about a young air force pilot who had bought a Tiger Moth and "didn't know anything about the aircraft except how to fly it" and was looking for someone with mechanical aptitude to accompany him to Edmonton and give him a hand if needed, Hal jumped at the chance.

Ready for a real adventure, Hal crawled into the back seat of the Tiger Moth, while the pilot jumped into the front where there was more heat. Hal wrapped a sleeping bag around his knees to stave off the freezing cold, and they flew out of Yellowknife bound for Hay River. Flying along the shoreline of Great Slave Lake, they came in on the west end of Hay River and landed.

Sleeping accommodation was the cabin floor of the fellow who operated a small charter and refuelling business. The next morning the intrepid flyers borrowed a heater to get the Moth started, and took off for Peace River.

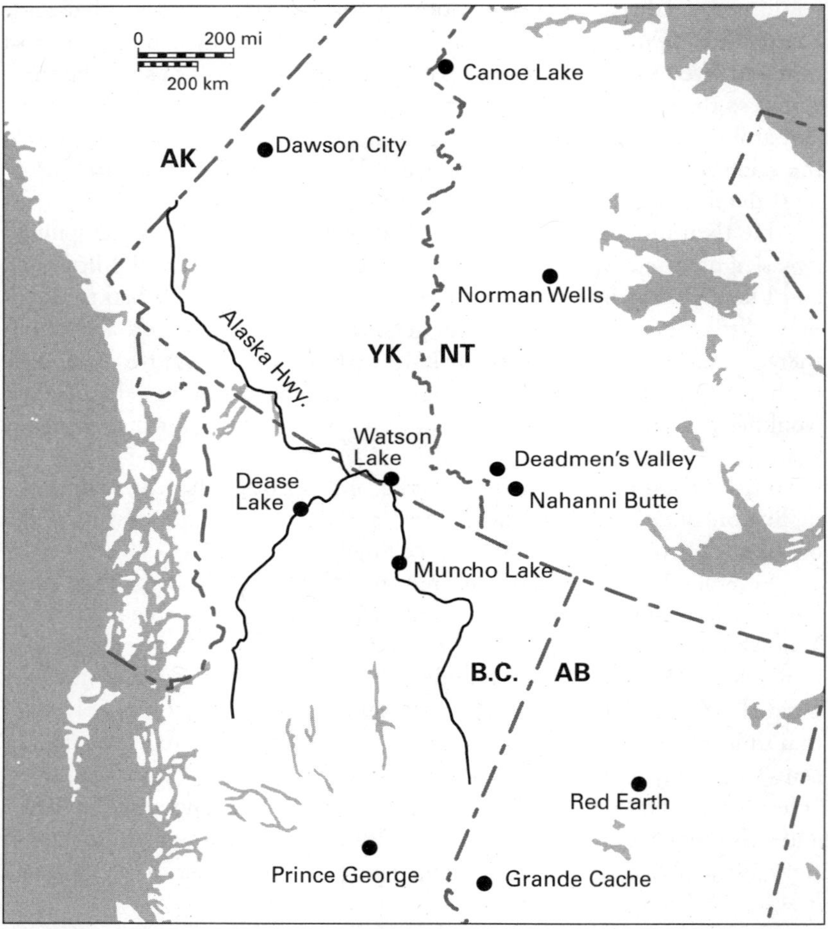

The "Rev" Hal Tetz: Helicopter Assignments

En route, they had to refuel, using some jerry cans they had aboard, so they came down on a slough, which they quickly discovered was blanketed with deep powder snow. As their skis sunk down, some twigs sticking up through the soft snow ripped the fabric on the underside of the lower wings. No matter, they had to get out of there. They flew on to Peace River, to the overnight luxury of a hotel. Should they carry on, ignoring the big tears in the wings and hope for the best? With the exuberance – or foolishness – of youth, they reasoned, "We flew it this far, we might as well continue." They took off for Edmonton – and landed safely.

"That trip in the Tiger Moth was my first ride with a pilot flying a privately owned aircraft," Hal says. "I was so cold, but it didn't matter. The weather was really bad so we were flying low the whole way, and with no navigation. I don't know how we found any of these places."

As soon as Hal turned 18, in 1945, he tried to join the air force, but his lack of education grounded that idea.

His quest for northern adventures satiated for the time being, he headed back home. After helping his father in the shop for a while, he took on jobs such as cutting meat in Ponoka in the cold storage plant, and working for Safeway in Edmonton.

Various jobs around southern Alberta came next: at Grassy Lake near Taber, hauling coal along the highway from a strip mine; at Barrier Lake driving a Euclid dump truck for Mannix Construction where a hydro-electric dam was being built; at Corbin in the Crowsnest Pass, hauling coal; at Lundbreck Falls, as a cat-skinner on a road-building project. That last job was especially memorable: "I had to drive up a steep rocky slope. Coming out on an angle the back wheel caught a rock and the scraper rolled on its side. We had quite a time getting that out of there. I got so much flack from the boss that I thought to myself, 'I haven't seen my parents for four years!' so I jumped on the bus and went to Vernon, B.C., where they had moved from Lacombe."

Another series of odd jobs followed: in a fruit packing plant, at a gas station.

In 1947, a heavy equipment operator told him there was good money to be made doing contract salvage work on the Canol Pipeline that had been built from 1942-44, and abandoned in 1945 after 13 months of operations. So Hal went north again – to the coldest job he'd been on yet. "It was unbelievable, trying to start up those old trucks and Cats. You couldn't touch a wrench outside with your bare hands or the frozen metal would burn you. We didn't have phones or radios – when you went out on the road you were on your own. Little did I know that many years later I'd be flying along there."

He journeyed south again and heard of good-paying jobs at Copper Mountain near Princeton, B.C., working underground as a nipper (miners' helper). "They were drilling a big raise. My job was to get all the steel, help set up and drill, get the dynamite ready and bring it up the slope, help load, and then put all the stuff away. Then we'd light it off and get out of there. They had two miners working together in the same raise, and I was a helper for both of them. They taught me so much about mining – you couldn't pay for lessons like that."

He next wrote his Explosives exam, and went drilling on his own. To gain experience in tunnelling he headed up the northwest coast to work in the Kemano tunnel, where "they did things in a big way." Water was being diverted from a dam on the Nechako River, where it came thundering through the mountain and down a huge 11-mile tunnel to the Kemano hydro generating station to supply power to the Alcan aluminum smelter at Kitimat.

He went back to Copper Mountain for a while, and then, considering the idea of taking on some overseas contracts, returned to Lacombe to open a bank account and establish an address. However, fate intervened when he met Betty Talbot working in the Alberta Treasury Branch. He was now 26 years of age, and it was time to settle down. He and Betty were married in 1953 – and the idea of a foreign mining contract was scrapped.

Instead, he found a job selling men's wear. The Jewish owner had a strong accent, and Hal's penchant for mimicking his manner of speech garnered him the nickname of "Rabbi". But the lower-paying jobs to be found in Lacombe no longer suited him, and when someone told him about the fun and profit to be made working in Banff driving tour buses, Hal checked it out and was hired by Brewsters to drive Greyline bus tours.

"We'd leave Banff in the morning, get to Jasper around suppertime, stay overnight, and make the return trip the next day," Hal described the work. "Then we'd have the next day off if they didn't have another tour."

Ever ambitious, and with any Seventh Day Adventist inhibitions long behind him, Hal filled in those spare days and evenings tending bar in the Cascade Hotel. Some of the Lacombe boys were also working in Banff, and when they came in to the Cascade and called for the "Rabbi" the nickname carried on. Then, through a series of comic events, the name changed to "Reverend", usually shortened to "Rev". Enjoying the joke, Hal would sometimes go around the bar with a tin plate asking for collections, and so the nickname stayed with him for life.

Betty moved to Banff and became the hotel's bookkeeper and assistant manager, and Hal eventually gave up the tour bus driving job to become the Cascade Hotel's bar manager.

But aviation was still his dream. He took flying lessons with the Calgary Flying Club to get his private licence, and luck was with him when he won a 50 percent interest in a car in a Kinsmen Club's raffle. He immediately sold his portion of the car, used the money to buy a half-interest in a Taylorcraft airplane, and built up his flying hours right in Banff in his own aircraft. He then went on to gain a commercial flying licence.

Helicopters

Fixed wing aircraft were all Hal ever thought of flying until the day Jim Davies, a helicopter pilot working for Bullock Wings and Rotors, landed in Banff. That changed everything.

When Jim offered to take Hal up for a ride and show him what could be done with rotary wings, he thought, "Man, this is the way to go! You can land anywhere you want!" Immediately following the initial helicopter ride with Jim Davies, Hal decided to get his helicopter pilot's licence, taking training first with Alpine Helicopters and completing his licence requirements with Bullock.

"At that time if you had a commercial pilot's licence you only needed 25 hours transition time to go from fixed wing to helicopter – which was really looking for a place to crash!" Hal says. "However, I didn't have enough time or experience, so one day in the spring of 1964 there I was walking across the tarmac at the old Chinook Flying School to see Neil Armstrong of Spartan Air Services about a job. He had airplanes and Hughes 269 helicopters flying up in the Arctic Islands.

"I was halfway across the tarmac when Gordie Bartsch of Range Airways came running up and asked if I wanted to fly right seat in a Canso. I said, 'I don't have a twin endorsement,' and he said, 'We'll get you one.'"

Flying Fixed Wing

Jimmy Lougheed (a cousin of former Alberta premier Peter Lougheed) had registered a charter for a Canso and two Norsemen under Bartsch's Range Airways licence. Four-and-a-half hours later Hal was taking a flight test to get his licence to fly twin-engine aircraft. Shortly afterwards he also took his heavy aircraft test. "It seems about all I did that spring was get endorsements."

So, in the summer of 1964 Hal found himself on the right seat of the amphibious Canso, with Jimmy Lougheed as captain (although Hal now had his captain's licence he never flew the Canso as captain), to transport sports fishermen from Edmonton to Great Bear Lake and back.

The first problem came with Hal not having float endorsement. That was okay leaving from Edmonton on wheels but they'd be landing on floats at Great Bear Lake. To appear legal while leaving Edmonton, they brought on board a Canso-endorsed Pacific Western Airlines [PWA] pilot to sit right seat. They dropped him off in Fort Smith, and then made the rest of the trip illegally with just Jimmy and Hal in the aircraft, to land on the water at Great Bear Lake.

Jimmy also had two Norsemen at Great Bear Lake used for side trips; there, Hal's lack of a float endorsement was easily fixed.

"We got into one of the Norsemen one night, no dual controls in it, I'd never been in a Norseman, but I did two landings. At that time a fellow with a float endorsement could sign out another fellow for a float endorsement, so Jimmy signed my licence." Hal never flew the Norseman again, however, keeping too busy on the Canso.

"I had a great summer and learned so much about flying that they don't teach you at a flying school. Bush flying is a little different; I really got the basics, where you had to make do with what you've got."

The Canso, "a big slow old bird," had been totally rebuilt for passengers, with rugs on the floor, proper airline seats, vinyl headliner, marine plywood side walls, and even a little biffy. "But," Hal adds, "you had to really *fly* that thing. It was hard work, and we didn't make much money in those days."

The routine involved weekly flights out of Edmonton to Branson's fishing lodge at Echo Bay on Great Bear Lake, with occasional side trips up to Coppermine [Kugluktuk], and to Tree River to fish for arctic char.

It was a busy summer. The pilots would load up at 11:00 at night at Branson's Lodge, stop to refuel at Yellowknife, and fly all night to arrive in Edmonton in the morning. They would then unload, refuel, take on new passengers and supplies, and fly north again, getting back the next day around 5:00 in the afternoon.

"Jim and I flew in 45-minute rotational shifts," Hal says. "The way the old Canso was built, it would drag along through the air at about 105 knots. Jim had learned if you trimmed the plane a bit nose up and then pushed forward on the stick, it was like getting a boat up on the step – it would pick up about 10 knots. It could just about keep up with a DC-3. But you had to adjust the seat so you were sitting back at just the right angle, and then lock your arms and hold them outright against the column. That's how we could rotate in those three-quarter-hour shifts. The person 'off duty' would just sit there in his seat and sleep."

Helicopter Hell

The summer job came to an end, and that fall, with his helicopter pilot's licence still unused, Hal received a phone call from Kelowna – a man who had leased an old Bell 47 from Anchorage wanted Hal to take it back up to Alaska.

Anxious to do some helicopter flying, Hal went out to look at the machine. "It was just a wreck," he says. "They had run the daylights out of it that summer and never did any maintenance. My first reaction was 'Suicide!' And it nearly was. But, I wasn't working for anyone, and I'd just got my licence. This guy needed a helicopter delivered up north, and to build up my hours I took on the job."

"It flew at 60 miles per hour," Hal recalls. "Later, I flew some of those old helicopters from Calgary to Banff, and would watch Volkswagens pass me on the highway! They'd get to Banff before I did."

Problems surfaced very quickly on this pioneering trip to Anchorage. Hal stopped to refuel in Kamloops and noticed that one of the long shaft hanger bearings had broken off.

"I had learned that every time you got out of one of those old helicopters you did a quick DI – daily inspection – to see if anything was wrong. They had those small control cables going out to the back end and you had to check to see if they were frayed or if something was happening. I sent for an Okanagan Helicopters engineer to come up. Then he had to send for the part, and wait for it to come from Vancouver. He welded it on, but then of course we had to line up all the cables. It took three days before I got out of there."

Hal and his Bell 47 resolutely headed north, loaded with gasoline kegs on one side, as he had to periodically put down to re-fuel with the wobble pump. "You always had to keep the balance. You, the pilot, were the heavy side sitting there, so you'd put a couple of kegs on the other side. However, you couldn't carry too many kegs because of the weight; either you wouldn't get off the ground, or you'd fly at 50 miles per hour until you burned the gas up!

"Luckily, though, those old things would take any kind of gas. I stopped at one of the service stations along the Alaska Highway, landed in the yard, and put in whatever av gas I had left in the tank. Then I filled the tank the rest of the way with car gas, and away I went."

He had more surprises ahead as he entered severe mountainous terrain. "Through the Toad River and Muncho Lake area the mountains aren't too bad, but as soon as you get past Whitehorse and up to the Alaska border, then you start hitting some big glaciers that come out to the highway. I also encountered ice fog as I got into glacier country."

As he and the old helicopter buffeted their way north and west up the Alaska Highway the weather became so bad that his flying time was reduced to an hour and a half a day. He crossed the Canada/U.S. border without any major incidents, and then hit a windstorm so harsh that all aircraft were alerted to land. He finally made it to the edge of the little town of Palmer and landed, staying overnight.

"The next day, although it was raining, I took off for Anchorage. But, going along in this rain, I didn't really know where I was at or where I was going. I knew the airstrip was by the water so I thought if I kept going this direction I'd eventually hit the ocean.

"I looked down and saw some numbers on the end of a runway, but I was so busy trying to fly with both hands that I couldn't tell what it was. When I finally did make it to the place where I was to deliver the machine, the fellow had some bad news for me. 'The Air Force wants you to call them.' I had apparently flown over a restricted USAF zone and I was ordered to come out and see them."

It was the last bad-news item that Hal needed to hear. "During that trip I had come down with the flu and I was deathly ill. So I told them, 'All I want to do is get into a hot bathtub and get some sleep. If I come out there now, half of you are going to be sick.' So the officer said, 'You can come out tomorrow.' I said 'Okay, no problem.'

"Then I turned to the guy that I'd delivered the machine to and said, 'You get me an airline ticket to Calgary, right now!' and the next day I was back home. I never went to see the air force people at all, and I never heard any more about it. And that was my first major trip."

Helicopters in Headless Valley

Hal's relationship with helicopters took a slight upward turn when, in the summer of 1965, Bullock Wings and Rotors offered him a job. It again took him north, to Muncho Lake and to the Nahanni area.

He flew from Calgary to Muncho Lake, "white-knuckled all the way," on summer contract for the Geological Survey of Canada. His assigned helicopter, a Bell Super G-2 with a blower on it to boost power, was, he says, the second worst machine he ever flew.

By the time he arrived at Muncho Lake, Mile 456 on the Alaska Highway and surrounded by 7,000-foot mountains, he had a total of 84 hours on a helicopter, with virtually no mountain experience except for his initial trip to Anchorage, and never with a passenger. At Muncho Lake he met with the brave company geologist he was assigned to transport around the area and very quickly learned how to navigate the "big rocks".

"But that machine!" Hal says, shaking his head at the memories of the old Bell. "At one point, we wound up parking it alongside the Alaska Highway and hitchhiking back to camp. They sent an engineer, who found that when they'd redone the engine they had put the oil scraper rings in upside-down. They'd made an *oil pump* out of it! You'd land and there would be fresh oil all down the tail boom and dripping onto the ground. So they changed that.

"Then, when I was way back in the mountains, the extension shaft on the tail rotor mid-bearing went out. I thought, 'I'm not staying here!' but when I tried to fly it out my feet vibrated so bad the soles of my shoes looked an inch thick. We made it to the Alaska Highway and shut it down there. They sent another guy up. He found out what was wrong with it and he changed that while it was sitting on the riverside."

In Fort Nelson he picked up Mel O'Reilly, the engineer that Bullock had assigned to work with him for the summer, and then flew into the

Nahanni region – famously called "Deadmen's Valley"*, or sometimes "Headless Valley" – to work with a geologist who had a summer contract. A local resident, Dick Turner, had been hired to bring in their supplies with his Maule, a high-powered, utility four-place tail-dragger aircraft suited to bush flying in rugged areas. Dick had carved out an airport behind his house along the foot of the mountain at Nahanni Butte, and basically taught himself to fly.

The men set up a bush camp and prepared for their summer sojourn in some of the wildest terrain in which to camp, explore, or fly.

"We were a long way from nowhere and we had no radios," Hal says. "Our supplies came in from Watson Lake once a week, and then Dick would come up there once or twice a week."

By the end of that summer, Hal was almost ready to give up flying. The very modest pay didn't help. "I still have a pay sheet from Bullock Wings and Rotors. Seasonal helicopter pilots made only $800 a month plus bush allowance, and a per diem for casual expenses of 50 cents."

It wasn't much of a return for a rugged existence away from home and family, and depending daily on his flying skills and the engineer's ability to keep the rotary wing aircraft aloft as they explored the crannies and crevices of Deadmen's Valley. But he survived, and so did his passengers.

Klondike Helicopters

Later that season Hal was hired as a pilot with a company that had better machines: Klondike Helicopters, and once again he was flying up North.

"My favourite customers were geologists," Hal says, "These guys were really good and if they found a pilot who was interested in geology they taught you about rocks and whatnot. They were the most interesting groups of people that I flew with."

During the winter months the company would bring all the machines back to Calgary, tear them apart and rebuild them. Hal thought that rather than do welding (one of many trades licences for which he had qualified earlier), he'd stay and work on getting his engineer's licence, which he did.

About that time, the Southern Alberta Institute of Technology invited four local helicopter engineers to advise them on necessary equipment parts and components, and specific courses, that would be needed to train apprentices for helicopter engineering. Hal Tetz, as one of the four, advised SAIT to ask helicopter companies to donate their time-expired parts for instructional use. The helicopter companies responded, especially Associated Helicopters in Edmonton, and enough parts were donated for the students

* See Chapter 1, "Deadmen's Valley", pages 18-30.

to build a complete machine. The result was the Bell 47G that is on display at the Aero Space Museum in Calgary.

Alpine Helicopters – Term #1

Hal was then hired by Alpine Helicopters, and he continued upgrading his qualifications to include his helicopter instructor's rating. Later, when Alpine sent him to run remote bases, he always went by himself because now he could do his own work.

Being qualified as a pilot and engineer allowed him to better understand both sides of the rotary business. "It was interesting to listen to a pilot coming in and trying to explain to an engineer what he'd experienced on the flight. Maybe he couldn't explain it clearly, or the engineer would interpret it a bit differently, but when you're flying it you get to know the machine, and after a while you know where to look to solve the problem."

While working for Alpine Helicopters, Hal was sent to fly a crew on a two-week job along the Yukon River up to Dawson City. On completion, the plan was for the crew to fly to Whitehorse and then on to Vancouver. However, the chief geologist first wished to take a look at another area.

Hal tells the story. "We put extra gas kegs on the racks of the Bell 47 and headed southwest toward Aishihik Lake. En route there was a small mountain of broken rock that looked like it had been man-made. We flew over to check it out, and saw that on the very top was a flat small patch of grass. The Chief Geologist asked if I could land on that. I replied that I thought we could – if we left the tail rotor sticking out into space.

"I got lined up into the wind and was quite high as in mountains you have to watch for the wind to come over the top of a knob such as this, and then descend real quick. You don't want to get into the downdraft. We sat down, and as the machine was winding down he asked me if I had noticed parts of a helicopter below the top. I hadn't, because I was too busy making sure we made a safe landing.

"When he got out and started to knock on rocks, I went over to the edge to see what he had seen. Well! Parts of an old G-2 helicopter were scattered around including some beat-up blades. I bent to pick up a small piece of wood from one of the blades and as I turned it over I saw an inscription written in black felt pen 'Here's to the G-2 – the underpowered son of a bitch!'

"It wasn't hard to figure out what had happened. The pilot had come in too low and got caught in the downdraft. I've seen this happen fairly often in remote places such as Ellesmere Island. It can happen anywhere," Hal concludes philosophically.

South East Asia Helicopters

Hal's wife Betty usually stayed home in Calgary to raise their two daughters, Cathy and Donna, born in 1963 and 1965 respectively. But in 1972 when he was offered a job with South East Asia Helicopters to go to Malaysia as a pilot/instructor, and had been there for a while to get oriented, Betty and the girls joined him.

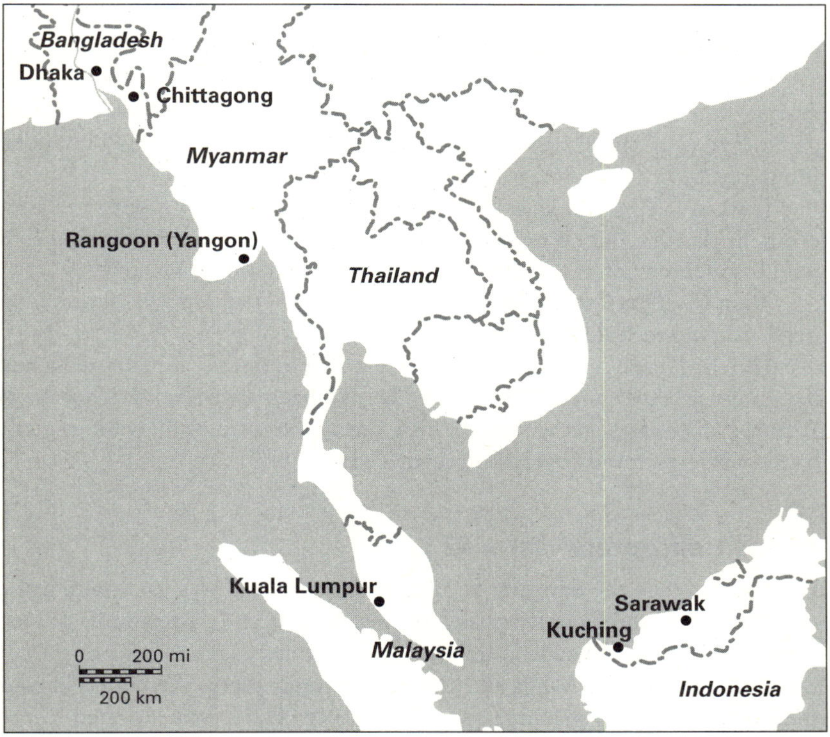

The "Rev" in Southeast Asia

In his role as the engineer he would accompany two pilots on flights with government personnel – "the highest rank, like the prime minister" – from Kuala Lumpur to Kuching, and out to villages in the jungle. Also, a short time after his arrival, while crossing the South China Sea en route to Sarawak on the island of Borneo, he checked out as a pilot on a Bell 206.

"I was there for six weeks when the company got a contract up in Bangladesh, so another fellow and I loaded up and flew there, through Thailand, up the coast to Rangoon [since renamed "Yangon"], Chittagong and Dhaka. I worked out of Dhaka, which is the capital of Bangladesh."

The civil war between East and West Pakistan ended with East Pakistan becoming the independent state of Bangladesh, on December 16, 1971.

"We were there just at the end of the war on a relief program, flying for the British government," Hal says. "They had a huge crane that floated on a barge. During the war, both ends of the railroad bridge had been blown up, and were hanging down into the water. A barge went there with a crane, lifted them out, put them on other barges and moved them to shore. I took the managers out in the morning, landed on a barge out on the river, stayed there all day and brought them back at night."

He stayed in Malaysia until, as he says, he darn near died. "I got some kind of a bug and had to quit flying.

"We had already enrolled the children in school, but one morning I was sitting there having breakfast and I asked the girls, 'How would you like to go back to your old school this fall?' I didn't get any argument, so I handed in my notice."

It wasn't so easy to leave, however. He first had to clear with the government that he had no outstanding records in their country and that he'd paid his income tax. It took two weeks to get everything finalized.

Then the Tetz family boarded a Thai International Airlines aircraft, and headed home to Calgary.

Hal had been in Malaysia six and a half months, and his wife and daughters just two months. But while he had been away, Jack Nicholson of Alpine Helicopters had written to him, stating, "Anytime you want to come back, put on your red coveralls and you've got a job!"

Alpine Helicopters - Term #2

Hal phoned Jack the day after landing in Calgary, and was told there was a mining rush in the Ross River area of the Northwest Territories. "I jumped in the machine and headed straight there. It turned out to be a bust, so I went over to Norman Wells and started a base for them. We built a hangar there and I had two machines working. Many oil rigs were located in the area, and we also did seismic, forestry, and environmental work, and flying work out on the old Canol Pipeline road, where I'd worked so many years ago." He came to know the people in this area quite well, and to admire their steadfastness in trying to make their lives better. "The priests up there had a pretty tough time making things go," he comments.

On March 21, 1974 the Canadian Parliament established an Inquiry headed by Justice Thomas Berger to review plans to build an oil and gas pipeline down the Mackenzie Valley. During the time Hal was flying for Alpine and operating the base at Norman Wells, he was hired to fly biologists around who were conducting studies for the Berger Inquiry.

"I did a lot of flying during the Berger Inquiry," Hal says. "Young people would come up doing surveys, but they didn't really have any idea what they were doing. Had they asked one of the old-timers they could have saved a lot of time and money. They had taken these classes, and, according to these classes, that was the way it should be …"

He warms to the subject, describing the antics of a group from a school in Toronto who came up one year to take mud and soil samples. "I was asked to sling a barrel each of gas and diesel out to them. I did, and then watched them dump it onto a grassy slope that drained down into the Mackenzie River. Then they marked out the flow with string on little sticks, apparently so they could record what damage such a spill might cause.

"When they came back the next year to tabulate this measurement, the snow and everything had slid down the hill, of course, taking their string and sticks and everything away. They couldn't find any sign of damage from what they had done the year before, so there they were, two years up there, and they'd learned nothing.

"I could go on for days about these useless programs and studies," Hal concludes.

Hal further recalls one late spring day when he was flying near Canoe Lake west of Aklavik and spotted two girls sitting out on a hillside. "I landed to see what they were doing, or if they were in trouble," he says. "The mosquitoes were literally eating them alive. Anyway, they told me they were from a university, and had permission to use a warden's cabin nearby as their base. There was one bank of snow left on the hillside, and they were taking temperature readings and measuring how fast this snow melted.

"When they told me this, I could only shake my head and leave them to their studies," Hal says. "Now, wouldn't that study be a big aid to the world, for somebody in South Africa, for example, to find out how fast snow melted up north? But these are the things that went on up there. I've seen biologists and other scientist types set up a camp on the ice, and then note that the ice was breaking up. I said, 'Sure the ice is breaking up – the weight of your camp is doing it!'"

Highland Helicopters

"Alpine was a great company," Hal says, "but they'd fight over a dime." Highland Helicopters paid better wages for a pilot/engineer, so Hal decided to make the switch.

In 1983-84, Highland sent him out with a Bell 206 to a base camp for Esso geologists near Dease Lake, B.C. There again he met unforgettable people, as independent and interesting as any he'd encountered in the North.

"Old Gerry Davis, down on Wheaton Creek near Dease Lake, had a Super Cub. He kept it out there all summer and flew around. It didn't have a radio or an Emergency Locator Transmitter, none of that stuff. Nobody knew where he went. That kind of thing went on everywhere."

Hal acknowledges that often aircraft maintenance in these isolated areas was not very good, mainly because there was nobody there to do it. "You had to do your own, and if you were a pilot you had to know how to fix that airplane. But things weren't maintained like they should have been, and a lot of it was like that."

Hal expresses his thoughts on such practices: "I worked for companies such as Alpine, Klondike, and Highland that had good maintenance, and anything that you asked for they would send you. Their idea was 'you'd better keep that machine going.' They were in a money-making business, and so were the pilots. We flew by the hour, so the more we flew the more we made."

After leaving Dease Lake, Hal came back to Alberta to start a base for Highland Helicopters at Grande Cache. "That was the best place I ever lived at a base. I loved those beautiful mountains and big plateaus, along the Rockies. It's just great country."

Hal next started a base for Highland at Red Earth, Alberta, between Utikuma Lake and Fort Vermilion, with two helicopters to service the oil industry. "We were really busy there. It was good."

By this time, Hal was considered an old hand at helicopter flying and fixing, especially in remote areas. He'd learned by rugged experience and also by listening to situations that fellow aviators had described and then tabulating their advice for future reference.

"A friend of mine once told me about flying a chopper along the coast of Tuktoyaktuk late one fall when his rotor blades began to gather ice and the collective started to come up. He knew it wasn't going to get any better, and he'd heard that if you pump the collective up and down as quick as you can that the blades will flex and you might get some ice to crack enough for it to fly off."

Hal explains that for increasing or decreasing lift, the angle of attack for all blades is "collectively" changed at the same time by the same amount. A pitch control lever called the "collective" is located on the left side of the pilot's seat and changes the pitch angle of the main rotor blades, resulting in climbs, descents, and changes in airspeed. To raise the machine off the ground and fly, you have to raise the collective according to what you want to do: hover or move into forward flight. A cyclic control, similar to a joystick, is located between the pilot's legs and changes the pitch of the rotor blades cyclically, and makes the helicopter move sideways, forward or backward. The tail rotor is an anti-torque unit: when the blades are going clockwise, the tail wants to go anti-clockwise. So if the tail rotor fails when you're flying, if you go down part way on the collective it releases the torque so it will line up again. But if you put it all the way down, then it wants to follow the blade.

"So when the collective started to come up one time, I thought about this pilot's advice," Hal says. "If you get ice on the blades the airfoil changes and the collective starts to climb, or come up. You have to try to hold it down – but sooner or later you must land. I first warned my passenger that I was going to pump the collective up and down so the blades would flex and hopefully get rid of the ice. He became very silent. I did this a few times and some ice must have come off because the collective wasn't as high as before, and was easier to hold down.

"However, the rotor system was now out of balance and it vibrated a lot. We made it to the oil rig we were going to and then it took two windshield scrapers and a small piece of pipe to crack the ice and get it cleaned up. Needless to say we took a different route back to Red Earth."

But different climates bring about different conditions. "Oddly enough I found that the opposite happens in places of monsoon rains," Hal continues. "When I was flying in Bangladesh, the raindrops were so big and heavy that I sometimes thought I was flying in a submarine. These large drops can pit the outer three or four feet of the blades. When the blades get too rough they lose lift. I actually had to file them down, like I sometimes had to do on the propeller of a float plane."

Animal Husbandry

Hal came to know people quite well in the various rural areas where he was asked to set up bases, and was often invited into local residents' homes for meals and visits. However, such welcoming gestures sometimes carried a secondary purpose.

"People would invite me to their homes for dinner, and this was very nice," Hal says, "until hunting season came along. Then there would be 'casual' inquiries if I'd seen any animals and where. I would give vague answers, but if I was paid to take out hunters I'd set them down where I knew they were at least a day away by foot from any animal."

He also encountered some paying customers whose ideas frankly disgusted him. "There was a biologist from Calgary who was doing a study for an oil company northeast of Grande Cache. We would fly out every day for a week so he could study the moose. Well, it's all forest covered and the only time you'd see a moose was in the early morning where they'd be on the sunny side of a ridge or something, just lying there. The biologist would want me to come over from the far side so he could spot them in an instant. But this scared the moose so badly that they'd go darting down the hill, sometimes end over end. I don't know how you could study anything doing that. I'm just not for it."

Hal recalls seeing a television show hosted by a well-known broadcast personality, about a group that had journeyed northwest of Old Crow near the Porcupine River in the Yukon to do a story on the Porcupine caribou herd. "The television documentary showed them chasing these caribou, running across the tundra as fast as they could, and then the announcer saying, 'This is their calving area.'

"Why, during calving season, would they be chasing the caribou?" Hal asks incredulously. "Not only that, but you could see the shadow of the helicopter on the snow as it hovered over the herd; you could also see the upper part of the frame, and on the TV you could see the tips of the rotor blades going, so I knew they were chasing them.

"A helicopter is a real indispensible machine, and so are fixed wing, and they both have prominent places in developing the north," he says. "But I don't go for that sort of thing, at all. Leave these animals alone! I know these guys didn't go out there purposely to harass them, but what they can tell from being up in a helicopter chasing an animal beats me. If they want to study them, get on the ground and camp with the animals and study them from there. I have seen animals chased until they are ready to drop, for somebody writing a book or going out there on a government grant and supposedly doing a study."

Nor does Hal believe in tagging bears or other animals. "Take a grizzly bear," he says. "When they meet a person walking across the tundra, they're going to take after them because they have a good memory and know that the last time they had anything to do with a human they'd got shot with a dart and when they woke up they had a horrible hangover. So they believe that person is going to do them harm and they will naturally try to do harm to the person first.

"I would say that the majority of these studies are a waste of time. If you were going to take a good look at an animal, and wanted to see what's going on, you'd be flying at about 100 feet, right alongside them. This is something that will traumatize any animal. I think it's so unnecessary. The things I had to do as far as wildlife is concerned during the Berger Inquiry, I was just shocked that people would do some of the things they did."

Changes

Hal often reflects on the places he's seen and the people he's met: the engineers and some of the pilots who taught him so much, how everyone helped each other out. "When we were out in the bush and you had to stand on a float to help guys do something, that's what you did. You helped the guy, and then you learned. You did things that soon become second nature, like walking. You 'feel' what's going on."

Although jokes prevail between fixed wing and helicopter pilots – "Helicopters don't fly – they're so ugly that the earth repels them!" – they are most often spoken with a smile, and accepted in the same light.

Hal saw very little evidence of pilots showing arrogance toward the engineers, especially in the helicopter business, because each was so dependent on the other. "Occasionally you'd find a fixed wing pilot who could walk on water," Hal says. "If they would start to razz me a bit about being a helicopter pilot, I'd jokingly tell them, 'Don't blame me for not upgrading your pilot's licence so you could fly one.' And that usually stopped them."

He says, however, that it wouldn't be of much interest for him to fly a modern airplane. "We did map reading. Now they've got GPS, they've got auto-pilots. It's not as interesting."

Aviation has changed over the years, as has the North Country in which he flew. "The way things are now, there aren't nearly as many people interested in getting into it," Hal says. "Young guys don't want to go out and live in a tent in a swamp someplace. And another thing is, the courses cost so much now that it's prohibitive. And you've got to have a minimum of 100 hours commercial.

"I really did love the North," Hal says. "The majority of the northern people, when I was up there, were super, particularly the old people, white and native. Everyone in the North would be interested in what you were doing, and wanted to help you. It was a whole different ball of wax than it is now with everything so government-controlled, and the fights going on up there over the ecology."

Hal cites the main things that changed the North were, ironically, "people trying to do good."

Helicopters 101

The "Rev" Hal Tetz has enjoyed the variety of aviation jobs he's held. Over the span of his career he amassed 11,226 total flying hours, of which 9,500 were on rotary, involving "lots of sweat time."

"I think you had to 'fly' more in those days than you do now," he says reflectively. "Nowadays you can 'power' yourself out of a lot of places. Back then you really had to fly an airplane or a helicopter; you had to do something in advance all the time. If you were getting ready to land, you had to know everything that was going to happen when you got to the ground. In the mountains there's no such thing as a sure bet, where you're going to land, or whatever. If you come in and all of a sudden a little whirlwind comes flying around the rock next to you it can create some real excitement. So you just had to think in advance."

He credits his knowledge, and his avoidance of serious accidents, to information gleaned from some very good pilots, such as Ted Jansen, who co-owned Alpine Helicopters with Jack Nicholson. "He probably taught me more than anyone, because they didn't want their machines banged up.

"You had to be on top of it, all the time," Hal says. "If you want to get to the top of a mountain you have to start climbing way in advance, like yesterday, unless you know the area and are familiar with it, and know where you are going to get an updraft to help you."

He explains that "a pilot would slowly climb, staying out of the downwind side of the mountain as much as possible, because you just didn't have the power to overcome some of those drafts. They could smash you right down. Not so much into the side of the mountain – that was usually a person flying along who wasn't really paying attention, and turning the wrong way – you always turn *away* from the mountain if you're going to turn around – things like that."

Some accidents occur because people are hurrying to get a job done. "A good friend of mine put a bunch of claim stakes in the cargo rack on one side of the helicopter, and then dropped off two prospectors and their rucksacks, and took off," Hal recalls. "The wooden mining claim stakes were on the rack on the same side as the pilot. They are heavy things and he had a whole pile of them, so when he let the passengers off, his C. of G. [Centre of Gravity] was off. He made a quick turn into the side of the mountain. That was the end of it."

Having "grown up in the industry" Hal can sadly recall many people and aircraft that have gone down. But, he says, many pilots escaped fatalities because they were in a helicopter rather than a fixed wing aircraft. "You might be settling into the trees, or into the water, but you can bring them to zero speed if your controls are still working. You're going down, you're losing altitude, but you're still flying them, you're still trying to keep it around 60 miles per hour."

There have been times, however, when a helicopter will spiral in if it loses the tail rotor and the pilot can't get on it quick enough to keep it straight.

"So," he concludes, "you've got to know what to do."

Bad weather was often the nemesis, but equipment failure caused Hal's one and only accident.

"I was flying a Bell G-2 for Klondike on August 29, 1967, and was ready to take off from Inuvik and bring the machine home," Hal recalls. "The helicopter had had some water in the gas line. I'd drained it but I guess it didn't all come out. I just got above the Mackenzie River when the machine popped and sputtered and split the needles, so the rotor and engine were going at two different speeds.

"If the needles are both together that means that they're locked in and the engine is driving the rotor. In auto-rotation, you split the needles; you put the needles to idle, and your rpm are indicated on one needle on the tachometer."

Hal further explains that old helicopters had a centrifugal clutch. While the engine was driving the rotor, the clutch was engaged. If the engine quit, then the clutch would release and split the needles, and the rotors would spin on their own as you descended in the air.

"I wasn't going to land in the water so I had to make it to shore right away. But the shoreline was awfully sloped, and before I could get turned right one skid caught the ground and I rolled upside down. That was the only incident in all those years, and in all those old machines."

Hal can offer a lot of advice – some commonplace, some extraordinary – to helicopter pilots or engineers who are working in isolated conditions with no help nearby, from memories from his own experiences.

"When you are both the pilot and engineer, and are working out on contract and have, perhaps, an upcoming 100 hour inspection, it can make for long days as you have to do the work at night after your day of flying. So my way of overcoming this problem was, I always carried tools with me. When I would be sitting out in the mountains, waiting for my clients to be picked up and taken back to camp, I would start a progressive inspection so I wouldn't have to do it all in one night." One time, however, he was doing some engineering work out in the wilderness on a Bell 206 when he encountered a bit of bad luck. "I dropped a bolt from the control tube on the head that the rotor blades attach to, and it bounced off the machine and into the long grass. I was sure I knew where it had landed but after a careful search I came up empty. However, I had an ace up my sleeve: I always carried divining rods with me in the machine so I got them out and did a search. I pinpointed the rods to a very small area, got down on my knees, moved the grass very carefully, and there it was.

"Now, not everyone can use divining rods, but I have the magnetism it takes," Hal adds. "I can also find buried wires, and water, or lost car keys, and – believe it or not, golf balls too – so I'm happy about that."

He offers to demonstrate this ability, noting that divining rods can be simply made even from coat hangers, and advises engineers working in the bush that they should never be without such basic equipment.

Setting it down

The "Rev" Hal feels quite blessed when he looks back over his career, sometimes flying less-than-topnotch machines through terrible weather and in isolated northern geographic locations.

"I always figured that God wasn't ready for me and the devil didn't want me," Hal laughs. "My project was trying to get people from A to B without any damage. I know a lot of pilots who would do a lot of things on the daredevil side, but I figured I needed the machine – and myself – for the next day."

On October 28, 1990, a Sunday afternoon, the "Rev" set down his machine for the last time and walked away. Since that day he has not so much as rented an aircraft to go for a spin for old time's sake. "I knew I might be tempted to go back."

Sometimes on an early spring morning he'll hear a machine going over, heading toward the mountains. It admittedly gives him a bit of a twinge, and he'll think, "Man, it's a beautiful day, he'll have a great flight!"

But then again he'll hear the slapping of rotary wings as a helicopter flies over on a cold, snowy winter's morning. He won't even be able to see the aircraft. Then he'll recall his own days of fighting wind and snow and rain in a bouncy little copter, hanging on for dear life amid the vibrations and vagaries of a machine trying to win over nature's wild elements.

"Oh, oh, he's not going to have a very good day!" Hal will say to himself. "I'm glad I'm down here!"

Kenn Borek Air – 'Anywhere! Anytime!'

When an airline such as Kenn Borek Air assures clients that it is willing to "fly anywhere, anytime, worldwide!" its employees and equipment must be prepared for some major adventures.

The added assurance of safe operations, advanced equipment, and the fastest response time imposes even more challenging demands. Therefore, employees know they might be called upon at any time to perform tasks that to others in the field would seem above and beyond the call of duty. And so, in 2008, when Sean Loutitt was named the company's Vice President, Operations, he was ready for new challenges.

Sean was never a stranger to extreme climatic conditions and geographic sites. His father, Alan Loutitt, both a pilot and an aircraft engineer, was working in Cambridge Bay, Northwest Territories (since renamed Iqaluktuuttiaq, Nunavut Territory) when Sean's mother became pregnant. There was (and still is) only a nursing station in Cambridge Bay, which was not set up for birthing, so Mrs. Loutitt was flown out prior to delivery. Further, Sean says "back in 1967 a pregnant woman wasn't supposed to fly 90 days before her term was up, so my mom came down to Edmonton and I was born there. We stayed in Edmonton for a while, and then went back north, to Inuvik and then to Fort Smith in the Northwest Territories."

For the next 15 years Alan Loutitt owned and operated Fort Smith Air Service, which at one time flew a dozen airplanes, and so Sean spent his early years immersed in aviation. "As soon as I could work I earned $2.52

an hour cleaning the bellies of the airplanes. I spent my summer holidays balanced on a piece of plywood on top of the float struts, lying underneath the aircraft and trying to come up with some concoction that would actually remove the exhaust from the paint!"

The Loutitts, like many aviation families, "bounced around quite a bit." They finally settled in Yellowknife, where Sean graduated from high school. "I'd always wanted to be a pilot, but then I became 16 and got my driver's licence, and it seemed that all that mattered then were cars – and girls."

Sean's father's advised him to broaden his education and not go straight into aviation, and so Sean attended the University of British Columbia to take a four-year degree in mechanical engineering. Following completion of his third year, he spent the summer working for Syncrude in Fort McMurray. From his downtown high-rise apartment, the view of float planes taking off and landing on the snye in the river stirred a longing to return to the field he had known and loved since childhood. "I went back to Vancouver, reduced my course load and started taking flying lessons. I got my private and commercial licences, then multi-engine, instructor rating, and instrument rating. I went through everything."

With a few courses still required to graduate, Sean worked as a flying instructor for the summer and then returned to university to complete his final semester. He graduated in 1990 with a degree in applied sciences.

Then, the aviation adventures began.

Sean's first job was with a small company called Spur Aviation in Yellowknife. "It was run by a man named Jensen, a mortician, who was married to Yvonne Quick. They used to call themselves 'The Quick and the Dead,'" Sean laughs. "I worked there for about four weeks and then got on with Buffalo Airways."

Buffalo Airways, a wide-ranging company with bases in Yellowknife and Hay River, NT, was (and still is) owned and operated by Joe McBryan, a friend of Sean's father. Alan Loutitt and Joe McBryan had started flying together some years earlier out of Fort Smith, and had shared a partnership in a Norseman aircraft. Joe now offered Sean a job with Buffalo Airways, working in the hangar with the engineers. He also helped Sean get his Pilot Proficiency Check [PPC] on the DC-3, which allowed him to fly revenue operations with passengers. It was a busy and exciting summer, and he did a lot of flying.

"Back then it was really hard to get a job, so it was definitely a tough working environment and Joe asks a lot of the new guys. It was a real apprenticeship into flying.

"Some of the guys I flew with, like Jim Smith, who passed away just a few years ago, were absolutely topnotch. I did my PPC the same day he did his, when he came to work for Buffalo as a captain," Sean continues, "and then probably 75 percent of my flying was with him. I learned a lot about those big radial engines on the DC-3, and how to load and unload the airplanes efficiently. We had to do it all. I saw a lot of the North over the three years I was there. By the time I left I was a captain on the DC-3, and a co-pilot on the DC-4."

Sean recognizes that modern-day Yellowknife is still one of the hubs of aviation. "Many people go up to get their first job and work for a bit, but all they can think of is getting back to Edmonton or Calgary or wherever. But really, if you're working in aviation in Yellowknife you're generally home every night, the money's good, and you get to fly fairly good equipment to very interesting places. It's actually one of the best places to work. And I had graduated from high school in Yellowknife so I didn't have any problems with living there."

After three years with Buffalo Airways, Sean took a short-term contract job with Air North. In the meantime he had married, but faced a tough commute with his wife being enrolled in a four-year degree college program in Kamloops in central B.C. The situation forced him to make some difficult decisions. He took a year off from flying and moved south, where he found engineering work in Vancouver.

Then in the summer of 1994 his father called him "back North", where he was now operating a small service called Reliance Air. Allan Loutitt had bought a second airplane, a Cessna 337, and asked Sean to join him for the summer to fly it. Following a season of flying out of Fort Smith with his dad, Sean returned to the lower mainland to sell one of the company's airplanes in Vancouver.

In the meantime, he had submitted his résumé to Kenn Borek Air Ltd., which had just bought a Super DC-3 and needed a crew. Because he'd had experience on that type of aircraft he was offered a job, and in September of 1994 he embarked on a career path that would change his life.

Go South, Young Man!

The twin engine Super DC-3 (DC-3S), purchased in Ontario and brought to Calgary, had been redesigned by Douglas in 1950. It was similar to the DC-3 but with a fuselage stretched 39 inches and other modifications such as newly designed wings with square-cut tips, an enlarged tail, and streamlined landing gear doors that fully enclosed the wheels. The Wright 1820-C9HE engines increased maximum speed over that of the DC-3 from 230 mph to 270 mph, and cruising speed from 207 mph to 251 mph. With more power it was capable of carrying a greater load, and seating up to 37 passengers.

Although Sean didn't yet have a PPC on this aircraft, because of his experience on the DC-3 the company hired him to fly the DC-3S. He was then given his first assignment: to fly it to the cold southern tip of Chile, and then continue to the Antarctic.

"We took off out of here and flew down to Kansas, Houston, the Caymans, Panama, Ecuador, and then from Concepción to Punta Arenas in Chile," Sean says. "We had engine issues from Concepción on. Every time we climbed above 7,000 feet one engine started running rough."

He explains, "There were two engineers on the aircraft and the first thing they changed was the carburetor, which was the most likely problem. Then they changed cylinders – but it just kept on running rough. They changed the sparkplugs, the ignition wiring, the magnetos. We were down there two weeks by this time, so finally they ordered a brand new carb. We got it and put it on.

"The next day we flew it and climbed above 7,000 feet. This engine was running great through to 10,000 feet – when the other engine quit! We feathered the prop, came back and landed. The engine had seized solid."

The engine they'd been working on seemed to be fixed, so they switched the engine on the left side with a spare they'd brought on board. They changed it, took it out, did a run-up and taxied back.

"The whole plan was that we were just going to fuel up, do a test flight, load and leave because we were trying to get down to the Antarctic," Sean says. "Well, when we shut down, the bearings seized on the engine we'd just installed! In the end, we ordered a new engine from the States. When it got there we had it changed, but by then too much of the season had passed so the contract was essentially cancelled. We loaded the airplane and flew back to Canada."

The problem required serious analysis. Losing an engine at any time is worrisome and in such a remote area doubly so, although the aircraft could be flown for a time on one engine. "I wouldn't say it was anyone's fault, because when they bought the airplane it had been in storage for a while and those engines were not meant to sit without being run. But it was very disappointing that we didn't get the airplane to the job."

Following this failed trip to the Antarctic in the DC-3S, Sean continued to fly wherever the company chose to send him.

"They sent me first to Nanaimo to fly a Beech 99, which seats 15 people and is similar to an unpressurized King Air," he says. "We were doing a sked [scheduled run] for Canadian Regional Airlines between Nanaimo and Vancouver."

He flew that route for a year, and then went up north for a summer to fly the DC-3 and Beech 99. He was then sent to Vancouver, and to Port Hardy on Vancouver Island, followed by another northern stint. Not exactly a pilot's dream, but nevertheless a job with a company he was coming to like very much, and he was building hours and experience.

In the spring of 1997 Sean was offered the position of Chief Pilot for Kenn Borek Air. Some major decisions had to be made at this point, both personal and professional. His marriage was breaking up, and the offer from the company occurred when many regional airlines were hiring, so finally there were lots of jobs available. Should he stay with Kenn Borek Air and take on the responsibilities of Chief Pilot, or should he apply elsewhere to fly big jets for a larger airline?

Sean had met the founder of the company, Kenn Borek, a few times and knew he ran a good company that was evolving in a positive way. Being somewhat influenced by his familiarity with smaller carriers from his earliest years of working and flying with his dad, he recognized the advantages of the company, and decided to stay. He was a bit surprised, however, to find himself instantly responsible for the management of 200 line pilots and their yearly training.

"I think the decision I made to stick around here, which is 12 years ago now, was the right one," Sean says. "I've been very happy doing all the stuff that Kenn Borek Air does. I really enjoy flying the Twin Otter, the DC-3, even the King Air.

"I've seen the company evolve from a bush company with no real SOPs [Standard Operating Procedures], all two-crew stuff, to the way we operate nowadays with the Twin Otters, the DC-3s, all the planes – more like the airlines operate their airplanes. Our pilots would have no problems going from Kenn Borek Air to Air Canada or to WestJet, because the environment they operate in here, in the cockpit, is very similar to how they'd operate once they got to these other airlines."

Home of the Twin Otters

In 1970, Kenn Borek of Dawson Creek, B.C. had incorporated Kenn Borek Air Ltd. as a subsidiary to his locally owned construction enterprise. Borek's experience with the petroleum industry in the High Arctic had taught him the rudiments of working in extreme climates: how to transport and operate a variety of machines in difficult terrain and remote locations; set up camps and maintenance sites; cache fuel and other supplies; rotate crews in and out – all in areas not particularly inviting to human habitation.

Borek's airside operation began with the acquisition of a DHC-6 Twin Otter (C-FABW). As equipment and personnel grew to meet project demands, the company positioned a number of permanent fixed-wing bases throughout Canada's northwest.

In B.C., the company flies out of Vancouver International Airport on behalf of Harbour Air, and has a fixed base at Sandspit in the Queen Charlotte Islands. Alberta bases are at the Calgary International Airport, and in Edmonton at both the international and municipal airports.

Northern Canadian fixed bases are located at Cambridge Bay, Resolute Bay, Rankin Inlet and Iqaluit, all in Nunavut; and at Inuvik in the Northwest Territories. In the spring and fall, barrels of fuel are brought north by ship and transported to caches along these air routes. During the flying season they will be picked up as needed from the bare "blue" ice, and hopefully not be covered over by snowdrifts. All maintenance and overhaul work is done in Calgary.

The financial centre for the Borek Group of Companies, where Rosella Borek is the CEO of Borek Construction, is in Dawson Creek, while the main office for Kenn Borek Air Ltd. is located in Calgary, headed as of this book's date of publication by President John Harmer.

World-wide operations for Kenn Borek Air are maintained on seven continents. Equipment and aircraft are leased to scheduled carriers, and air support services provided, for jobs throughout Canada, in Asia and South America, and the Maldive Islands in the Indian Ocean. The company transports scientific teams and adventure charters in both north and south polar regions.

Kenn Borek Air has also undertaken tours of duty for United Nations humanitarian relief and peacekeeping missions in Africa, Pakistan, Mauritania and other countries. As well, the company sometimes provides people and equipment to various other concerns that fly throughout the world.

The Calgary base of Kenn Borek Air is the No. 4 Hangar built in 1940 for the British Commonwealth Air Training Plan. It is sturdy and original. Stepping over the raised door sill, a visitor enters a busy reception area that leads to a labyrinth of hallways and to small no-frills offices.

The wood-frame hangar has high-ceilings and is painted sparkling white, with a compact but tidy floor space. The place is a beehive of activity. Experienced trades people work at welding, hydraulic, sheet metal, aircraft refinishing, and float manufacturing. The deHavilland DHC-6 Twin Otters painted in the white, red and black company colours chosen by Kenn's wife, Rosella, sit with tails up and noses down to receive thorough servicing before continuing their duties as airborne workhorses.

These stalwart aircraft provide the basis for Kenn Borek Air operations. Forty-two of its 60-plus aircraft are DHC-6 Twin Otters, making it the largest Twin Otter operator in the world. The aircraft's general reliability, as well as its short landing and takeoff abilities and adaptability to wheels, skis, or floats, make it an ideal aircraft to operate in variable terrain. With the company holding domestic and international licences to provide charter and scheduled services for all locations of operation, employees might find themselves flying anywhere from the two Poles to the tropics. An example of one of their hot-spot locations is the Republic of Maldives, where one can find "a lot of water, and a lot of Twin Otters."

The Maldive Islands are comprised of 1,200 coral atolls, or islands, scattered across the equator. "Each island is tiny, like one city block big," Sean says. "One island will have one resort on it and the only way people could get there, originally, was by boat. Visitors would have to sit on these boats for three or four hours in chopping seas to get to a resort where they're spending thousands of dollars a day. So of course you realize there must be better ways of doing it.

"A Russian company was running some old military helicopters and crashed a few, so then they eventually brought in a Twin Otter. Now Kenn Borek Air supplies everything – Twin Otters, pilots and engineers – on contract to Maldivian Air Taxi. Right now we have 22 Twin Otters there, all on floats, and we move a couple thousand passengers a day. That's truly a Twin Otter airline, the way they operate.

"We work five or six days a week, and it's usually 32 to 35 degrees Celsius outside so in the cockpit it's 45 degrees. There's sweat rolling down your arms, and you're doing 20 legs a day, in short little hops. There's one trip that's two miles, like, a minute."

"But," he adds, "There aren't a lot of jobs where you can work in shorts and a shirt, bare feet, and fly a $2 million float plane around all day. But, it's a *lot* of hard work."

On the opposite side of the flying field, some pilots and engineers choose the coldest and most remote climatic zones, in the Canadian Arctic or the Antarctic. "But usually we get people coming in who say, 'I want to go everywhere!'"

He adds, though, that "everything gets you away from home, which is tough, whether it's the Antarctic or the Maldives. We get some people who are happy just to rotate up into northern Canada, and some who only want to work out of Calgary and fly up to Fort McMurray, or to B.C. – to Kamloops, Vancouver, Campbell River – or to Saskatchewan or Manitoba, across to Toronto or Halifax, or down to Houston. We fly wherever people want to go."

The Antarctic Ice!

With its matchless experience with the Twin Otter, and with a superbly qualified crew that proudly upheld the company's motto of "anywhere, anytime", Kenn Borek Air was the logical choice to perform an extraordinary feat: to make a mid-winter medical evacuation (medevac) flight into the cold, dark Antarctic, where the odds against successfully completing the assignment were a bookie's dream. And thus in April 2001 Sean Loutitt's extraordinary adventures began.

The destination was the Amundsen-Scott South Pole Station in Antarctica, the southernmost continually inhabited outpost on the planet.

Named for Roald Amundsen, who reached the South Pole in December 1911; and Robert F. Scott, who reached this destination a month later, this U.S. research station was originally constructed in 1956 and rebuilt in 1975 to house researchers, supplies and equipment in a geodesic dome and connected outbuildings. Other structures were installed in the 1990s to serve astronomy and astrophysics work being done there. The snow-surfaced Jack F. Paulus Skiway at the station, 12,000 feet in length and 9,300 feet above sea level, handles all air traffic. Up to the 1998-99 summer season (October to February) the U.S. Navy serviced the site with LC-130 Hercules aircraft. From that time on, and only during the summer (when the temperature might reach a balmy 7° F), the responsibility for daily cargo and passenger flights to the station has been handled by the New York Air National Guard's 109th Airlift Wing.

While the station is populated by about 200 personnel in the summer, most depart by the middle of February, leaving anywhere from 25 to 50 scientists and construction workers to "winter over", with their only means of communication being Internet emails and satellite phones. It is an isolated spot, and extremely dangerous throughout the winter, with temperatures hovering around -75° F or, with the wind-chill factor, -143° F. The annual mean temperature is -56° F. The skies are pitch-black for 24 hours of the day throughout the winter. Snow and ice cover the terrain. The station is self-sufficient, powered by three generators running on JP-8 jet fuel. A physician stays through the winter to look after medical concerns – that is, until the physician *himself* becomes the concern.

When the resident physician at the Amundsen-Scott South Pole Station, Dr. Ronald Shemenski, developed acute pancreatitis after a gallstone plugged a duct between his pancreas and gall bladder, it was clear that he would have to be brought out in order to save his life. Another physician, Dr. Betty Carlisle, who was familiar with the site and was willing to return, could be brought in to look after the station's residents throughout the rest of the winter season. If the trip could not be made immediately before winter completely closed in, the ill man would have to wait there until October, six months away. The situation was untenable.

First, three USAF National Guard LC-130 Hercules aircraft had been called to Christchurch, New Zealand, to check out the situation and formulate a plan. Their best shot involved multiple supply drops that would require large numbers of aircrew and ground personnel. Also, they'd need 2,000 gallons of heated fuel in order to get back from the Pole. However, it was reasoned that the Hercules couldn't come in – or have any hope of getting out again – because their hydraulic systems could be unreliable in temperatures colder than -58° F, which in fact was a highly probable scenario. Although the LC-130s are the backbone of the U.S. Antarctic program, which continually operates large aircraft in and out of glaciers and ice strips at high altitudes, two rescue attempts were made and thwarted because of weather, until the mission was called "impossible".

With time and opportunity quickly slipping by, Raytheon Polar Services Company of Centennial, Colorado, which holds the contract with the National Science Foundation to provide services for these Antarctic bases, now suggested that the U.S. government turn to the only other airline, and aircraft, in the world that might be able to make the trip to the South Pole: Canada's Kenn Borek Air. Raytheon knew that this airline, which pledged to fly "anywhere, anytime," was staffed by "bush pilots and engineers" who were knowledgeable about flying in extreme cold temperatures and darkness, and who operated a fleet of ever-reliable "bush planes", deHavilland DHC-6 Twin Otters.

The emergency call came in to Kenn Borek Air in April 2001. Stephen Penikett, then Operations Manager, and Sean Loutitt, Chief Pilot, were called to consider the risks. By then, Sean had made five trips to the Antarctic and logged over 2,000 hours in deHavilland Twin Otters.

If Kenn Borek Air accepted the challenge it would be the first time in history for an aircraft to go on a mission "to the bottom of the world" when weather conditions were so extreme and unpredictable, and then land and take off in total darkness.

"I wasn't going to go on a suicide mission," Sean states. "We had to examine all the possibilities. We didn't want to all of a sudden be blindsided by something unknown, so we tried to think of anything that could happen so that we were fully prepared for the flight, to completely minimize the risk."

The Twin Otter could be totally relied upon to do the job. "Kenn Borek built this company around the Twin Otter, because it was able to go out in the middle of nowhere, where he wanted to build roads or service oilfields. And today we still pick the Twin Otter because it can handle stuff and go anywhere," Sean says. "We didn't require any extra engines. The Twin Otter has pretty good engines; we trusted those engines.

"The lack of moving parts makes the Twin Otter work so well: fixed gear, limited hydraulic systems, initial flight testing at colder temps; deHavilland didn't put a temp limit in the manual, therefore each operator can determine its own limitation. LC-130s have a temp limit in the manual and so no one is allowed to exceed that limit."

While both the Twin Otter and the LC-130 have excellent short takeoff and landing [STOL] characteristics, the Twin Otters are smaller and lighter and therefore require less length on the skiway. And, while the LC-130s do regularly go to the Pole, sometimes four times a day in the summer, for this mission other factors needed to be considered. For this extraordinary journey, less skiway would need to be prepared for the Twin Otter (not all 12,000 feet), which would lessen the burden on the South Pole workers when making the all-important preparations for the aircraft to make a safe landing and takeoff. Also, in the event that the flight, once at the South Pole, was unable to take off, the Twin Otter carrying four extra people (two pilots, an engineer, and the nurse) would be easier to accommodate for six months than the 20-some that would accompany the LC-130. With limited resources at the Pole, this was a very important consideration.

Two Twin Otters would be equipped for polar navigation and long-range operation in cold weather, and fitted with wheel-skis. Most of the airplanes that go to far northern or southern regions have permanently installed engine heaters, de-icing equipment and global positioning systems, but for this journey extra features were added such as ferry tanks for additional fuel, Inertial Navigation Systems [INS], and safety and cold weather equipment including triple-layer sleeping bags, flares, snowshoes, shovels – and life rafts just for the ferry flight in case of an ocean ditching. An iridium satellite phone system guaranteed an open line of communications regardless of situation. The loads came to 17,500 pounds in each aircraft, including 7,000 pounds of fuel. At this weight, the usual takeoff run of 1,000 feet would rise to over 5,000 feet, but this weight and fuel load were necessary for the journey ahead.

The flight was expected to take five days to complete the 7,600 nautical miles [8,745 statute miles] from Calgary to Punta Arenas, a port city on the cold southern tip of Chile, and then another 800 nautical miles to the British Antarctic Survey base at Rothera on the Antarctic Peninsula. One reserve aircraft would stay at the Rothera Station in case a search and rescue response was needed, while the other would go on to make the 1,346-nautical-mile flight, requiring an estimated 9½ hours, to the South Pole. Essentially, the entire flight from Rothera to the South Pole would be in darkness, as Rothera itself was only experiencing about an hour of daylight at that time.

Choosing the right people for the job was the next task. On April 9, Sean phoned Captain Mark Cary, who was currently flying for the company in the Arctic. Sean described the mission and asked, "Are you up for the trip?" "I'm ready," was the eager response. Mark then contacted his apostolic pastor to advise him of this upcoming challenge, and receive the pastor's blessing and prayers for a safe and successful mission.

Aircraft Maintenance Engineer Norman Wong was called next. He hesitated ever so slightly, knowing the incredible responsibility that would rest on his shoulders to keep the machinery operating in such extreme conditions. But then he too said "Yes." This crew, captained by Sean Loutitt, would be the one to make the last, crucial, leg of the pioneering journey in to the dark regions of the South Pole.

Of the three, Sean was the only married man. Sean's wife Sandi, whom he'd met on the job at Kenn Borek Air and married in 2000, was also a

Twin Otter pilot. She was well aware of the risks, but also understood the incredible excitement connected with the challenge. CTV's "W-5" program that covered the mission quoted Sandi Loutitt's calm statement, "You just plan the mission. You do your best on the ground and in the air. And then, if anything happened – I would have to come and rescue him!"

The second Twin Otter would come as far as Rothera Station, crewed by Captain Matthew Gacek, First Officer Anthony Szekely and Aircraft Maintenance Engineer Peter Brown.

"Everyone who went down had flown in the Antarctic before, they were all experienced pilots, coming and going from Rothera in the South Pole," Sean says. "I think it was a very well-put-together plan. And besides just doing the trip, it was great to be part of the whole planning and execution – it wasn't like just getting your assignment and doing it. We came up with the plan and then we got to execute it, which made it much more exciting."

Their gravest concern was, of course, the cold temperatures at the Pole. They had to be prepared for anything down to -103° F – *nothing* could be left to chance. "So you're looking at grease, you're looking at the temperature that fuel has to be, you're looking at avionics, aircraft instruments, tanks, how you're going to plug in the engines, how you get extension cords out there," Sean says. "The area is in total darkness and there are no runway lights at the South Pole, so you have to figure out how you're going to get adequate lighting to land in 24-hour darkness."

And then they went for it.

The Journey into Darkness

The two Twin Otters left Calgary at 5:00 am on April 14 to begin the long five-day journey into the night-dark skies of the Antarctic, flying south through the United States, Central and South America, and on to Punta Arenas. They landed there on April 18, 2001, and departed at 6:45 am EDT on April 21 to fly 800 nautical miles over the choppy waters of the Drake Passage [Mar de Hoce] and to Rothera Station.

They landed safely in Rothera that same day and made final preparations for the arduous journey ahead. But the weather turned even worse than imagined, dropping to -90° F. While they waited for clearer conditions, the two engineers decided to "toss coins – best of three" to see who would go with the mission plane. Norman Wong won this dubious honour, to be the person responsible for the ski-equipped aircraft as it flew through darkness to land in heretofore untested conditions.

Meanwhile, early on Tuesday, April 24, amidst bad weather, a Royal New Zealand Air Force C-130 plane reported that it had safely lifted off from the airfield at McMurdo Station, Antarctica, a U.S. base on the Ross

Sea about 850 miles from the South Pole, after rescuing 11 Americans from a Raytheon Polar Services outpost there.

The replacement medical staff came on board: Dr. Betty Carlisle, who had agreed to stay at the Amundsen-Scott South Pole Station for the winter; and a nurse, Betty Erickson, who would accompany the patient on the return medevac. So when a 36-hour period of flyable weather was forecast that day, the Kenn Borek Twin Otter, captained by Sean Louitt and Mark Cary, with engineer Norman Wong, Dr. Carlisle and Betty Erickson, took off at 10:34 am EDT on April 24. Their 10-hour, 1,346-nautical-mile flight into utter darkness, and cold that could snap steel and turn liquid fuel to jelly, had begun.

Three hours later, they viewed the remaining horizontal line of an orange sunset slipping away behind the aircraft. "The sky is amazingly clear," Sean Loutitt reported, as he talked to the U.S. base at McMurdo and other Antarctic stations during the trip. "You wouldn't believe the number of stars we can see."

The moon was new, which meant there was no moonlight, and no auroras lit the sky. This considerably helped communications. For long distance flights such as those in Antarctica, high-frequency radio bands allow clear communications, but this frequency can be blocked out by solar disturbances that create auroras. With no auroras, "the coms" were amazing. They flew along at 10,000 feet at an average speed of 145 knots through this winter wonderland, enjoying the moment, and trying not to focus too heavily on the difficulties that might be experienced when they came in to land.

Meanwhile, the folks at the Amundsen-Scott South Pole Station were having their own challenges. The airstrip, the Jack F. Paulus Skiway, was located over a mile from the station. The staff brought out tractor Cats and ran chains across the ice airstrip to smooth it off, while fervently hoping that the steel tracks on the Cat wouldn't break from the extreme cold. The Skiway was too far from the base and it was just too cold to attempt to connect electric lights to outline the strip, and electric power cords would surely break in pieces unless wrapped in heat tape.

The aircraft couldn't "land blind" onto the pitch-black icy surface of the airstrip, so the only solution was to bring out a number of metal 45-gallon burning barrels and fill them with wood debris and gasoline. But the gas turned to crystals, refusing to light in the high winds and -91° F temperatures! A final solution was to bring out propane "Tiger torches" to warm the material and set it alight.

Figuring out how to bring the aircraft in safely would solve only 50 percent of the dilemma. How would they keep 800 gallons of fuel heated for the takeoff? And would the metal components on the aircraft snap when started again? First things first.

Back at the Calgary base, Steve Penikett was sweating bullets as he listened to the clear communications being sent by the airborne crew. He turned anxiously to the media reporters gathered around. "The wind's blowing like hell!" he announced. "We're getting reduced visibility and blowing snow."

Six hours into the flight and 15 minutes before reaching the "point of no return" where they would no longer have sufficient reserve fuel to turn back if the weather turned bad, the crew held a quiet and serious discussion about possibilities and probabilities.

Then Mark turned to the captain, "Sean, this is it. We're gonna do it."

"Yes, you're right," Sean replied, as Norman nodded agreement.

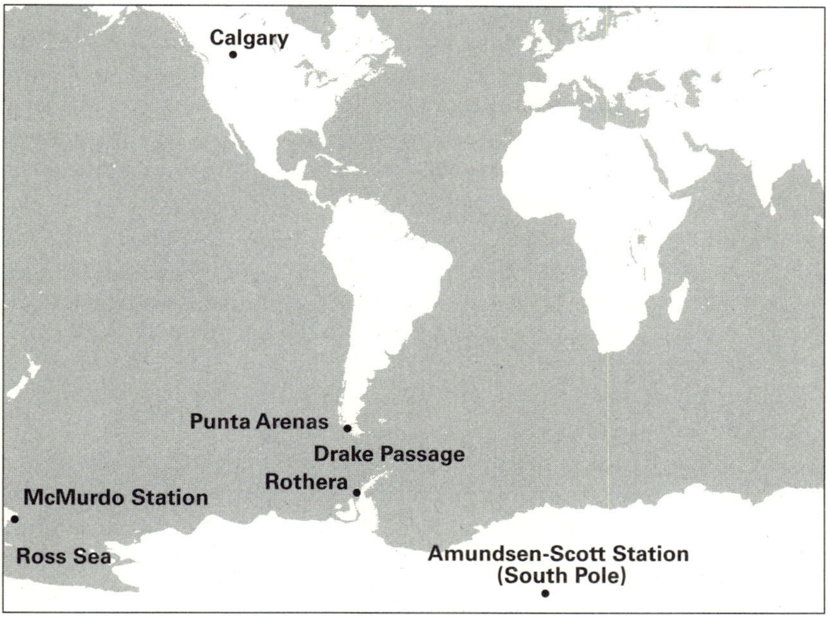

From Calgary to the South Pole

And then the weather worsened. Swirling ice crystals severely reduced visibility from the cockpit as they came in for a landing. The pilots and engineer stared intently through the crusted windshield in vain attempts to spot flames flickering from 22 burning barrels lining the airstrip. But suddenly, there they were! In they came: 100 feet above ground level, 90 feet, 80 feet, 70 feet, 60 feet.

"A lot of emotion was going on," Mark Cary recalled in an understatement. Power was reduced at touchdown and suddenly the exterior conditions became interior concerns when the cabin heat ceased. The cabin is heated by "bleed air" from the engines, and even in -40° temperatures it can become very cold when the engines are at idle. But no matter – they'd made it, with a safe landing at Amundsen-Scott South Pole Station located at 90° south, at 8:02 pm EDT, in total darkness, with the temperature at -90.4° F. The 1,346-nautical-mile flight from Rothera had taken 9.3 hours.

"And that's how we managed to make the world's first landing in total darkness in Antarctica," Mark Cary says. "We had arrived at the bottom of the world."

"When I came in to the South Pole station, the bottoms of my boots were a solid block of ice," Sean Loutitt recalls, "yet my feet were still warm. We were dressed for these conditions."

Because of the difficulty in getting the material in the barrels to light, the ground crew kept them burning for the entire 12 hours that the Twin Otter remained there. The crew ate and slept, while the aircraft and the equipment inside it, as well as the fuel, oil and hydraulics, were kept warm by electric cords wrapped in heat tape to connect the heaters. Tents were draped over the engines as soon as they were shut down, to be warmed again before takeoff with portable Herman Nelson hot-air blowers. Batteries were removed and stored in heated trailers.

"All airplanes have systems to keep the fuel warm," Sean says. "Modern jets fly above the tropopause* where it is -50 degrees for every flight. We had the fuel heated at the Pole because we were worried that the oil might not get warm enough to heat the fuel prior to takeoff, and with the engines operating at max, the fuel consumption is quite high. Once the engine is operating at high settings the oil will be warm and therefore the internal system should work properly."

Sean and the crew had high praise for the work undertaken by the ground support people. Because of their careful preparation of the airstrip and concern over the lighting and accommodation, the air crew knew what to expect. "Making this flight onto a runway lit by burning barrels of debris, with the temperature dropping to 90 below zero, wasn't nearly as challenging as my summer flying into the Antarctic," Sean said later. "Sometimes, while taking scientists to set up remote camps, what looks like solid snow can hide crevasses. The onus is on yourself and the flight crew to use judgement."

Following the required 10 hours of rest came the work of getting the aircraft into the air again, with the ill doctor and the nurse as anxious passengers.

Space and comfort were secondary to safely getting up and away. The two 950-litre ferry tanks sat lengthwise in the cabin, forward and aft of one

---

* The level separating the troposphere and the stratosphere, occurring at an altitude of five to 10 miles.

another. Their positions were, essentially, directly above the main internal tanks in the belly of the aircraft and so did not affect the centre of gravity – as long as proper burn-out was maintained from each tank. Since the plane was not pressurized it was drafty inside the cabin; at least it was warm, however: "between +10° C and +15° C – the Herman Nelson heaters really did their job," Sean says.

The passengers and engineer sat in the far back for takeoffs and landings. For the return flight, an area on top of a pile of gear was cleared off so the patient, Dr. Ron Shemenski, could lie down. An intercom system allowed the pilots to converse with those in the back, and vice versa, but at any time any person could climb over the gear and come up to the front.

Takeoff would be another challenge. Engine and battery heaters were plugged in, and gas-powered blowers fanned heated air inside the aircraft and through ducts to the engines and hydraulic bay compartments. It temporarily became so warm inside that Sean, at the controls in his flight suit, had to remove his head-gear.

But even with all these precautions, trouble started immediately. The controls wouldn't move. Norman Wong and Mark Cary removed the ceiling panels to find the flaps frozen in an extended position, which would make it impossible to complete the flight. Of all the flight controls on the Twin Otter, only the flaps were hydraulically operated.

"There weren't too many options but to shut the aircraft off and do some urgent repairs," Norman said later. Their other choice would be to remain at the South Pole for six full months, with the continually ailing doctor and very tense staff.

"You develop a sort of heightened awareness," the engineer said, referring to his analysis of the problem and the subsequent repair work that rested fully on his shoulders. "And so you go back to 'bush maintenance'. This basically meant that you do whatever had to be done to get going, in this case cutting the cables, jerry-rigging the controls and getting the flaps fixed."

That done, a new problem surfaced. The aircraft skis had become stuck fast to the ice and the plane could not move! Although wheel skis had been put on in Punta Arenas, they were a hydraulically operated "bucket" type, so it was not advisable for the system to swing the bucket with all the aircraft weight still on it – and definitely not something one could do at those temperatures.

So, again, bush maintenance was called upon. A D-6 Cat was driven to a position just under a wingtip, and Mark Cary climbed on top of the Cat to show "two big men" how to gently – and without damage – rock the fiberglass wingtip to free the skis. Sean stayed at the controls to keep the engine power set up.

Slowly, slowly the aircraft began to move. "It was one of the slowest, longest takeoffs in Twin Otter history," Mark recalled – a full 7,000 feet were required to gain sufficient takeoff speed. And so, after five hours of preparation, the Twin Otter departed Amundsen-Scott Station at 12:47 pm EDT on April 25, while the air crew silently wondered if the instruments could take this beating from the cold.

"The main flight instruments are gyros. They have to spin very, very fast – and *nothing* spins fast at those temperatures! Since there is no horizon to enable a pilot to visually orient the aircraft in relation to the earth, and without any lights on the ground to help determine land from sky, the only way to know whether you are straight and level was the attitude indicator. In the same way, the altimeter just reads static pressure from outside the airplane, but if there is a blockage of the static port – an unheated passageway that directs static air to the airspeed indicator, altimeter and vertical speed indicator – you might think you are at a safe altitude but you are not. Again, there are no lights on the ground to give you any visual clues," Sean says.

So, could the instruments really be trusted to operate properly in such conditions? Were they headed in the right direction?

"The challenge with flying and navigating from the South Pole is that every direction leading away from it will take you due north," Mark Cary explains. "We had to be certain that when we departed the South Pole we were heading north to Rothera and the South American continent, as opposed to Africa or Australia. The global positioning systems installed on the aircraft enabled us to fly to any point on the earth with pinpoint accuracy. We put this system to the test!"

On they flew through the dark sky – to be rewarded by a slight pink line appearing on the horizon. "What a gift!" Mark said later.

Sean agrees. "It was a nice warm feeling, going toward the land of light."

The pioneering Twin Otter landed safely at Rothera at 8:52 pm EDT. Then both the Kenn Borek Air Twin Otters departed Rothera the next day at 11:20 am EDT, with Dr. Shemenski and Nurse Erickson, and arrived 4½ hours later, at 3:55 pm EDT, at Punta Arenas. There, Sean faced the most uncomfortable time of the entire trip, when they were met by the media. "I was colder at the press conference in Punta Arenas on Friday morning following our return than I was at the South Pole. I hadn't dressed for an outdoor conference!"

Back in Canada, excitement had hit a fever pitch. The little Canadian aviation company that promised to fly anywhere, anytime, and its brave and competent crews had for the first time in world history brought the South Pole out of total winter isolation.

That fall, Captain Sean Loutitt, First Officer Mark Cary and Aircraft Maintenance Engineer Norman Wong were presented with a Meritorious Service Decoration by Canada's Governor General, The Right Honourable

Adrienne Clarkson. The civil division of this honour recognizes individuals who have performed an exceptional deed or activity that has brought honour to their community or to Canada. The citation for their 2001 mission to the Antarctic states "… The success of this mission is directly attributed to the professionalism of Messrs. Loutitt, Cary and Wong, who brought great credit to Canada, both nationally and internationally."

Further awards followed for the crews of the two Twin Otters who made the pioneering voyages, including the 2002 Trans-Canada (McKee) trophy for outstanding achievements in the field of air operations, the Flight Safety Foundation Heroism Award, and the Canadian Owners and Pilots Association [COPA] Special Award of Merit.

An irony occurred when so much media and film documentary attention was given to the U.S. counterparts and the work of the LC-130 Hercules and crews that regularly fly to the South Pole to service the U.S. Antarctic Program stations, and virtually none to acknowledge the Canadian team and the trusty deHavilland Twin Otters that had made this epic journey. Even today, calendars and postcards depicting the Antarctic reveal fantastic scenes of frigid beauty – with no photos of a Twin Otter that brought the photographers to these isolated and nearly inaccessible sites.

However, the game was won, and best of all, the near-fatally-ill patient, Dr. Ronald Shemenski, recovered his health following a cardiac operation (when it was discovered he'd also suffered a mild heart attack), and later had gall bladder surgery. He then became director of a medical facility in Denver.

While the year 2002 brought fame to Kenn Borek Air, it also brought incredible shock and sadness. The company's founder, Kenn Borek, and his daughter Carleen Borek-Walker lost their lives in a traffic accident, leaving his widow Rosella, his family, and a caring staff to carry on the traditions of excellence that Kenn had sought to establish.

Retracing the Flight

The following Antarctic summer, Kenn Borek Air flew back to the area to complete their regular season. Both Sean and Sandi Loutitt piloted two of the aircraft working there for the season.

"Sandi has flown as a captain down to the Antarctic, she's flown in the Maldives – she's been around the world," Sean says appreciatively. "She knows what we do in this company, so it is nice because she does understand all the evolutions."

(Sandi flew for Kenn Borek Air until 2008, and is currently an Embraer First Officer for Air Canada.)

In September of 2003, Kenn Borek Air was again contracted by Raytheon to undertake another rescue mission to Antarctica. After being informed that the New York Air National Guard's LC-130 Hercules transport aircraft again could not make the trip in the cold weather, two Kenn Borek Air Twin Otters left Calgary one day apart on Monday, September 8 and Tuesday, September 9, 2003. The first aircraft was captained by Sean Loutitt, with First Officer Brian Crocker and Aircraft Maintenance Engineer Kevin Riehl; the second, auxiliary, aircraft was captained by Jim Haffey, with First Officer Rob Forbes and Engineer Shawn Erickson.

The last regular flight from the South Pole had occurred in February, with the next flight not expected until October, but here they were, mid-season, and fighting weather again. The first aircraft arrived at Punta Arenas on Friday, September 12. The expectation of taking off on Saturday, September 13 for the six-hour flight over the Drake Passage to the British Rothera Base on the Antarctic Peninsula was delayed because of weather. They did get away the next day, however, with the second aircraft arriving a day later.

The temperature, recorded at -70° F on September 18, was a bit kinder than it had been on the previous medevac in 2001, but winds gusting up to 50 mph with blowing snow kept them from proceeding further. The sun does not rise at the South Pole until September 21, and the area was now in what is termed "civil twilight" for 24 hours a day, which barely gave residents at the base enough light for outdoor activities without using artificial illumination.

The patient, and the reason for this medevac, was Barry McCue, an Environmental Health and Safety officer employed at the station by Raytheon. The resident physician at the station, Dr. Will Silva, had consulted with the RPSC medical director Ronald Shemenski – the same Dr. Shemenski who himself had been rescued by the Kenn Borek Air crew in 2001 – and it was determined that McCue required immediate gall bladder surgery. Evacuation was highly advisable and it would have to be done as quickly as possible because the condition could only worsen.

Depending on the weather, McCue might have to be transported to the U.S. base at McMurdo station on the Ross Sea, south of New Zealand, instead of Rothera. In that eventuality, the National Science Foundation had requested, through the New Zealand Antarctic Program, to have LC-130 Hercules aircraft from the Royal New Zealand Air Force 40 Squadron stand by to make a flight to McMurdo to meet the Twin Otter.

The Kenn Borek Air rescue flight departed Rothera at 8:48 am EDT on Saturday, September 21 to fly 1,346 nautical miles from Rothera, and arrived at the Amundsen-Scott South Pole Station at 5:30 pm EDT that afternoon. The 58 resident scientists and service crew had been busy preparing a welcome, which again consisted of constructing smudge pots to guide in the aircraft, and grooming the runway in the -60° F weather. After landing, bamboo poles were spread underneath the skis so the aircraft would not stick to the ice and prevent takeoff, the engines were again blanketed with insulated covers, heaters were used liberally to warm the engines, and the batteries were brought inside.

Not only did the Calgary aircrew bring emergency relief to the patient, but they also transported 100 pounds of tomatoes, avocados and oranges – the first fresh food the inhabitants of the base had had since running out of such supplies in April.

Following a 12-hour layover, the crew with their patient departed the South Pole station at 5:00 am on Sunday, September 20 – the day before "official sunrise" – in overcast skies with the temperature recorded at -67° F. Although not dark, clouds prevented them from actually seeing the sun. Before takeoff the aircraft circled around on the icy airstrip to warm the skis, so only half the runway length was required to get airborne.

The Twin Otter arrived at Rothera at 1:51 pm that same day, where the patient changed planes and departed for a flight to Punta Arenas. From there he was transported back to the United States on a Lear jet chartered from a Chilean ambulance service. Surgery was successfully performed on September 25 at the University of Texas medical branch in Galveston. Another life saved.

Referring to former mid-winter landings and takeoffs that had successfully been made to and from the Amundsen-Scott South Pole station, Barry McCue lightly stated that in order to come for him, "they [the Kenn Borek Air crew] just needed to take the plan off the shelf, blow the dust off, and then just figure out what the people should do." But the very grateful patient effusively praised the crew during his interview with the Los Angeles *Times*, "Those guys are great. They had just flown 10 hours through the roughest environment known to man."

The rescue Twin Otter landed in Punta Arenas at 10:30 pm local time on Monday, September 22, and both aircraft and crews returned home to Calgary on Wednesday, September 24 – with "mission never impossible" accomplished once again.

Creative Challenges

In 2008, following the retirement of Steve Penikett, a 30-year veteran of the company, Sean Loutitt was named Vice President, Operations for Kenn Borek Air. He acknowledges that it's a big job but he's "into it."

"It's always tough balancing everything, but I guess trying to evolve the company into something better is a goal," Sean says reflectively. "There are so many people who have so much potential in this company – the company is built around the people performing their jobs. The rescues we've done, and all the work we do in the High Arctic, the South Pole, and the Maldives, it's all a tribute to the people who work here."

Future goals are, basically, "always trying to be the best at what we do, which is essentially to provide specialized services."

The Twin Otter remains a vital component of the varied assignments undertaken by Kenn Borek Air. With a total of 844 deHavilland Twin Otters that were manufactured, the first in 1965 and the last in 1988, the aircraft that are presently owned and operated by the company are continually being rebuilt in the Kenn Borek Air hangar in Calgary. "When they come out of our shop I'd say they're as good as or better than new," Sean says.

However, Viking Air Limited of Victoria, B.C. is bringing the Twin Otter back into production. Viking secured the design rights from Bombardier (which had acquired them when they purchased deHavilland), along with other types such as the Dash 7, Buffalo and Caribou tactical airlifters, Otter and Beaver bush planes, and the Chipmunk trainer. Now a new line of Twin Otter, called Series 400, will give new generations of air crew the versatility to go "anywhere, anytime".

"Viking building them will be a new chapter for the Twin Otter," Sean says happily.

Along with their 42 Twin Otters, Kenn Borek Air operates two turbine DC-3s. This turbo-prop conversion, manufactured in Oshkosh, Wisconsin, modernizes the DC-3 so it can operate from unimproved strips as well as on skis from snow surfaces, and thus complement the versatility of the Twin Otter. Kenn Borek Air now flies these DC-3s in the High Arctic as well as in the Antarctic.

"I'd say the Twin Otter and these DC-3s are our 'backbone' aircraft," Sean says. "Then for support airplanes we have Beech 99s, and King Air 100s and 200s for our northern bases, which give us the ability to service communities.

"The Twin Otter is really a specialized airplane. If you want to go from, say, Resolute Bay to Cambridge Bay, it's more efficient to do it in a King Air than in a Twin Otter, but if you want to go out to an esker somewhere in the middle of nowhere, the only option is a Twin Otter. The local base managers will assign each trip, consulting with me, and we'll pick which airplane would best suit a particular base. That's how we assign aircraft, and then the chief pilot will ensure that a properly trained crew is there."

Now that Sean Loutitt's aviation dreams have been mostly fulfilled, he is perhaps in a position to offer advice to others who are thinking of coming into the business. He considers for a moment what he should say.

"Well, everyone has heard the advice before, which is, 'it's always tough.' I heard that advice before I became a pilot. Family life is always hard, so people have to think carefully about what it is they want to do. Do they want to follow the crowd, or are they actually trying to do something that they *enjoy* doing?

"You really have to love flying," he emphasizes.

But such a career is not for everyone, especially for those who yearn for the conveniences of a city and a comfortable work routine. To build a career in aviation, a person has to get used to being away from home, and moving to different places to get a better job or more experience. The going can be tough, but the tough get going, as the saying goes.

"There are many people working in the industry who love their jobs right up until the day they retire – yet you see others who are unhappy at what they're doing, whether they're flying for Kenn Borek Air, or for WestJet or Air Canada," Sean says. "But they get in a groove, and think, 'Well I can't leave now because of seniority …' or for whatever reason. I think a lot of people start out with false aspirations; they think it's so wonderful until they actually get there. And then they say, 'Now I just want to make money so I can go and do other things, because this is not really that great.'"

A retired helicopter pilot, reflecting on a career that involved being away from home for up to six months at a time, stated that not many young people want to pursue career paths that once were followed as a matter of course: working as dock rats, fuelling, cleaning, loading and unloading aircraft, living in camps or tents in remote northern locations, flying and fixing in tough environments.

"I think there's definitely something there," Sean says. "People do want that instant gratification – and the helicopter industry in particular is difficult. I almost went and became a helicopter pilot after I was a fixed wing pilot. The funny thing was, I was just unhappy about not being near home until I realized that the helicopter guys usually have it worse."

Sean tries to not work too long hours but it doesn't always pan out that way, especially with his new position of VP, Operations.

"Right now this job is new and it is definitely all-encompassing. When I became Chief Pilot it was a similar situation. You've got to invest some time to set the system up so that things are working smoother. You need to spend some time developing ideas and building client bases, taking care of customers and stuff like that. You've got to have some good people in key positions so you don't overload yourself. We have excellent people like Peter Brown, our Maintenance Manager, and Corey Marchuk as Safety Management System Manager, Tony Szekely as Chief Pilot, Deborah Kominko as Assistant Chief Pilot, and others."

An important part of the job, Sean says, involves spending some time each year at all the company's bases, to fly with the various crews and meet the people. "Just sitting behind a desk doesn't allow management to know what's really happening out there. I've been at all the bases so I have a pretty good idea, but I also know that things change year after year. You can lose touch with what's happening."

There's a lot to keep track of: 60-some aircraft, including 42 Twin Otters, two DC-3Ts, 16 Beechcraft King Air 90s, 100s, and 200s; a Beech 99 called the Airliner and two Embraer Bandierante ["Bandit"] E110s that fly south to Antarctica or north to Inuvik, Rankin Inlet, and Resolute, and to the Canadian West Coast, across Canada, and to the Maldives.

But what, really, is the main challenge with the job, as one tries to carry on the traditions set by Kenn Borek Air? Sean considers his response, and then says thoughtfully, "Dealing with people, motivating people.

"I want everyone to like Kenn Borek Air as much as I like Kenn Borek Air. I want all our employees to want to be here. But like I said, they're a great bunch of people; they have invested in the company and we want to keep it that way.

"Right now, of course, I have so many new challenges …"

Even greater than flying into -90° F darkness?

He nods, and opens the door from his office that leads into the reception area. Phones ring incessantly, and people crowd the entranceway to "have a minute" with the Ops VP.

Outside on this winter day, piles of snow are banked around the crowded parking lot of Kenn Borek Air's Calgary hangar. The air feels cold at -26° C, but the sun shines and the sky is a bright Alberta blue. That's about as serious as it gets in Calgary's winter season.

But at each tip of the world the winter season brings total darkness, blowing snow, and temperatures so cold that steel can snap and splinter like kindling wood. However, as has been proven, nothing can prevent competent crews, flying the trademark white-red-and-black Twin Otters, from successfully penetrating the world's darkest and most forbidding domains.

'So Others May Live'

"I first learned of CASARA's existence in 1986 when 13 people lost their lives up there in the Kananaskis country," Garry Wutzke says. "That's when I got the desire to get involved with the organization, to help search for people who are missing."

The inspiration for Garry's involvement with the Civil Air Search and Rescue Association unfolded like a Greek tragedy: first, on Friday, June 6, 1986, a Cessna 182 leased from Hughes Aircraft was reported missing in the south-western Alberta region of the Rocky Mountains known as the Kananaskis. The two people aboard were pilot Kenneth Wolff from Ponoka, and Fish and Wildlife biologist Orval Pall from Calgary, who was conducting field wildlife studies.

The next day, friends Bruce Pratt, Roderick Harradence and Hayden Evans went searching for them in a Cessna 185. They too went missing amidst turbulent weather: the deadly combination of rain clouds, high wind, and snow cover in the high Rockies.

Searches were immediately put into effect, and the news media actively reported every piece of information they could find regarding the missing personnel, their aircraft, and the terrain that was now being searched by up to 16 aircraft and ground searchers.

"There were thunderstorms in that valley and most likely that's what put the first aircraft down," Garry reflects. "Then friends decided to go out in this thunderstorm area and look for them. Friends' impulsion to find friends. Unfortunately they encountered the same situation."

Search master Major Mike Barbeau of 440 Squadron came in to direct air searches that covered over 6,000 square kilometres. Rescue officer Lloyd Gallagher, Emergency Services Supervisor, Kananaskis Country, acted as ground crew co-ordinator when over 100 Kananaskis Park Rangers plus personnel from the RCMP, Fish and Wildlife Department and Alberta Wilderness Association as well as friends were brought in to assist. All battled extreme obstacles of weather and geography that blocked discovery in this desperate race against time to find the men alive. But no traces of sheared-off treetops or flash forest fires or smells of burning oil or smoke were detected. Nothing.

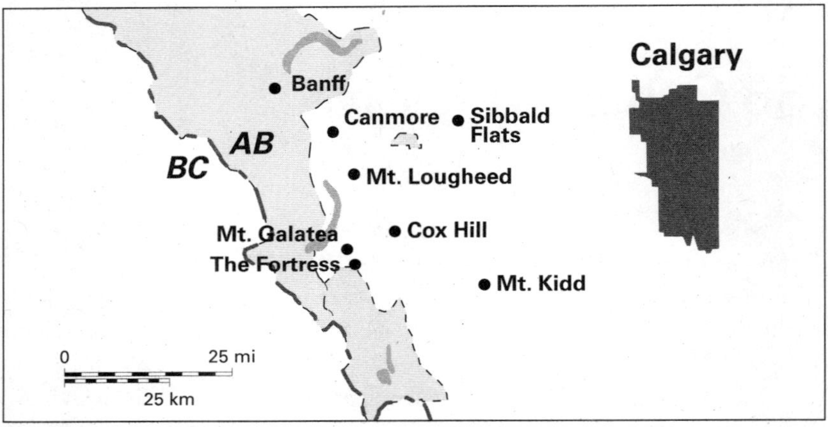

Greek Tragedy in the Kananaskis

On June 9, an aircraft wing section was spotted at the 7,500-foot level of the snow-covered northeast face of Mount Lougheed. It was the second aircraft, the Cessna 185. As Jim Lapinski of Aero Copter Services in Cochrane described, retrieval was impossible due to extreme turbulence that grounded further air search or rescue operations. But ground searchers continued to clamber through the bush and climb impossibly steep and difficult rocky terrain in hopes of seeing a sign indicating the whereabouts of the initial missing aircraft, the Cessna 182.

On June 10, further military assistance was requested, and on June 12, Day 7, the Combat Convoy from CFB Calgary joined the search with 150 soldiers (110 searchers and 40 in the support group) and 36 trucks to transport the ground camp equipment. They set up a base on Sibbald Flats campground and began to make arduous, organized treks through 500 square kilometres of dense bush and rocks east of Mount Allan and the Barrier Lake region.

Then on June 14, Day 9 of the search, a third aircraft went missing! The military de Havilland Twin Otter was carrying Captain Donald Wayne

Plumtree and Captain Edward (Ted) Kates of 418 Air Reserve Squadron, Edmonton; Sergeant Brian Burkitt of 4 Regular Support Unit, Edmonton; plus five searchers from Civil Air Rescue Emergency Services or CARES (later renamed Civil Air Search and Rescue Association, or CASARA): Carl Grant, David Hall, Charles Masur, Patricia McLean and Jerome Schindler. The wreckage of the Twin Otter was discovered by Alberta Forest Service personnel when they spotted a fire at the 5,500-foot level of Cox Hill in the Kananaskis, which then extinguished itself in the snow-covered area to leave only a scorched patch of terrain. The military flew in a Hercules aircraft and personnel to perform the sad duty of retrieving bodies, while Major Wally West of 440 Squadron, Forces Base Edmonton brought in soldiers from Princess Patricia's Canadian Light Infantry [PPCLI] to guard the Twin Otter crash site until all debris could be removed.

"They lost their lives trying to save others," declared the CTV news anchor.

That statement had a profound effect on Garry Wutzke. "The investigation showed that when the accident occurred there was a shadow effect," Garry says. "The mountain they hit in the Twin Otter was in the shadow of another mountain. They didn't see it until the last moment and just ploughed right in."

On June 16, Day 11, the search was called off for the Cessna 182, setting off personal and political furors. Due to pressure to resume the search from then-Alberta premier Don Getty and Ottawa politicians such as Member of Parliament Harvie Andre, the search was restarted the following day.

A military Challenger jet brought Colonel Ray Brown from Regina to head the inquiry, and the formal Aviation Safety Board investigation commenced.

By June 18, Day 13, monetary costs were adding up at an alarming rate. Two military Twin Otters continued to search at a quoted flying cost of $2,000 per hour, and choppers at $1,200 per hour, plus meals and accommodation for all personnel.

On June 19, Day 14 of the search, helicopter pilot Jim Lapinski and spotter Clay West sighted the white tail section of the Cessna 182 at the 7,000-foot level on the west side of Mount Kidd, near The Fortress and Mount Galatea and not far from the Kananaskis Golf Course. The bodies of Kenneth Wolff and Orval Pall were retrieved the next day. The ordeal was nearly over, at least for the tired but dedicated searchers.

Pilot Jim Davies came in with a helicopter and 100-foot steel cable and nets to perform the dangerous but necessary task of hauling out the wreckage so investigators might discover the cause of this tragedy. The searches had lasted from June 6 to June 20, and those at home watching the news updates on television could now find some relief from the tension.

Three mountain lakes in the Kananaskis Country were dedicated in memory of the 13 lost lives (henceforth known as Memorial Lakes), with the public dedication held at the Mount Kidd Park Amphitheatre on September 27, 1986. Tributes were made by military, government and civilian officials, with Pilot Officer John G. Magee Jr.'s well-known poem "High Flight" recited by Justice A. Milton Harradence, QC – the grieving father of one of the deceased and himself a pilot. For many, the heartache resulting from this triple aircraft tragedy that took 13 lives would never end.

As the story faded from the public eye, Garry Wutzke's thoughts galvanized: he was an experienced flyer; perhaps he could do something to help in these situations.

"It's amazing how accidents like this get you involved," he says. "You'd think a person would shy away from that type of thing, yet it seems to get people more interested. In my time with CASARA, I've found that our best membership increase comes after a major search. Once things are quiet and there's no searching going on, people seem to find other things to do. It seems there's a need to solve these problems, and it does get people involved."

The Flying Farmers

The Wutzke farm was west of Skiff near Foremost in southern Alberta, where fields are large and weed spraying (or crop dusting) could be taken care of quickly and easily – but perhaps not always safely – by taking a few flying lessons, buying a small aircraft, and doing it yourself.

"My older brother Melton took lessons and got his licence first, at the Lethbridge Flying Club," Garry says. "Then we bought a Cessna 120 and put a sprayer on it. I took flying lessons in Calgary at Chinook Flying Service when I was 20 years old, in 1960.

"It sounds pretty dangerous," he acknowledges, "Most people would say, 'You've got to be crazy!' but we went out and sprayed a lot of crops with that Cessna 120."

The dangers came mainly from "power" – and by that he means not only power lines, but also the lack of power in the Cessna 120.

"Generally we'd do our spraying in the morning, but sometimes you have to finish the farm fields and so you keep going. It can get quite hot later in the day and it's difficult to get off the ground on a hot day, so we were going under the power lines. We had to spray under the lines anyway, but the thing was, we didn't have enough power to get over the lines until we got enough speed up. Once we got going it was okay.

"And then we got a Cessna 175, followed by a Cessna 180. We sprayed our own and the neighbours' crops – not recommended for living a long life and seeing your grandchildren – but it did the job for the time being."

While there is a law restricting private pilots from doing spray work outside a 25-mile circumference of their own farm, Garry notes "There's a lot of land within 25 miles, so that wasn't a problem."

A small accident happened when Melton hit a power line that sheared the top of the tail off the Cessna 180. "During the heat of the day the power lines would sag quite a bit," Garry says, "and when you're crop dusting you don't look at the power line, you look at the power poles. Unfortunately the line had sagged quite low. Just about all Cessnas have a rudder nose [an 'aerodynamic balance' used to reduce the control loads] that points forward. The rudder is a lot heavier than the rudder nose, so it's pointing forward and acts like a balancing point. As the line slipped over the rudder, it caught the top of the tail and ripped six inches off it, sharp, like with a knife. He just kept on flying and finished spraying that year with the broken tail."

The main problem was that with the balancing point gone, the rudders became very hard to push. But Garry just shrugs. "No problem other than that, and we learned to compensate for it."

Their parents must often have wondered if their sons' flying was a good idea. "It's amazing how they suffered through it!" Garry says, but it helped with the crops and brought in a little extra money. Most farmers take on at least a part-time job during the off-season to keep things going and subsidize the farm. The Wutzke boys brought in this extra money during the spring and summer season, with the added bonus of the enjoyment of flying.

Crop dusting became a seasonal ritual even after Garry moved to Calgary to take a diesel mechanic's course at the Southern Alberta Institute of Technology [SAIT]. As his career moved ahead and he had to put in long hours to gain accreditation for the four-year course, crop spraying took a back seat. "Melton continued with it for quite a while, but when he acquired a wife and small children and thought of the potential dangers of the job, he decided he should perhaps quit while he was still alive. He didn't stop flying, but he did stop crop spraying."

Garry started his employment with Mack Trucks, and then Trimac, followed by CP Transport. The longest term, which lasted 23 years, was heavy duty mechanic and later shop foreman for Alberta White Trucks. Next came the position of service manager at Freightliner Trucks "until the stress of upper management got too heavy," and he retired. However, that was cut short when he was called back, and in 2009 he is happy to be "still going to work."

Garry stopped flying for a few years when his children were young. "I was then making $2 an hour at my job, and with small kids, flying would have been difficult!" But a few years later his wife Margaret presented him with a gift certificate for a flight with Chinook Flying Service at Springbank Airport west of Calgary.

After two hours of dual, he was hooked once again. For the next couple of years he rented airplanes, generally Cessna 172s. When a neighbour decided to sell his Taylorcraft and upgrade and suggested that they buy an airplane together, they purchased a Cessna 172. Garry flew it for five years, adding 500 hours to his air time, but soon he wanted something bigger, faster, and more powerful. In 1983, he teamed up with his brother Melton to buy a Cessna 182, and added 1,700 more flying hours on that aircraft.

Then, in 1986, came the accident that changed many lives forever, including that of Garry Wutzke.

Searching for Life

Shortly after the triple crashes in the Kananaskis country, Garry decided to attend a CASARA general meeting to learn more about the organization. At that time the 200 or so local members held their meetings at SAIT. There, Garry learned that all civil aviation members across Canada volunteered their services, with the only paid position being an administrator in Winnipeg. Members were reimbursed through grants from the Department of National Defence and Transport Canada for out-of-pocket expenses when training and when flying on searches, based on the price of fuel and the horsepower of the member's aircraft engine. If members were required to fly a long distance on a search, the organization also paid for meals and accommodation on a per diem basis.

The national organization was subdivided by province and territory that functioned under the same standard operating procedures. Military Rescue Co-ordination Centres [RCCs] were responsible for initiating all search and rescue activities. Each would contact the civil air search and rescue operations that came within its jurisdiction, and would also be responsible for evacuating ill or injured people from remote locations to major hospitals. Alberta zones responded to commands issued through 435 Squadron in Winnipeg.

When a zone commander was alerted to prepare to mobilize a search operation, information and briefings would be issued, and appropriate air traffic control centres notified to maintain a listening watch for the missing aircraft. If this phase did not result in discovery, the "distress phase" would come into play, requiring a full search and rescue operation: plotting the flight route on a map, determining probabilities, issuing a Missing Aircraft Notification, and appointing a search-master and assistants. When the aircraft was located, resources best suited to conduct a rescue were notified, as well as the Transportation Safety Board and next of kin.

CASARA in 2009 lists 375 aircraft and nearly 2,600 volunteers in provincial and territorial zones across Canada, and becomes involved not only in searches for missing aircraft but also for anyone who is deemed to have missed their Estimated Time of Arrival [ETA] at their destination, be it from road, water, or air travel, and includes persons such as hikers or hunters who have gone missing in the wilderness.

Following attendance at that seminal meeting, Garry joined the organization and took their training programs. He has now, in 2009, been a CASARA pilot for 20 years, and has risen through the ranks to become Commander of Alberta Zone Two, which includes all members from the Southern Alberta area. "There is no Zone One," Garry says. "It used to be Lethbridge, but now Zone Two looks after everything south of Calgary to the U.S. border, and on the west and east to the B.C. and Saskatchewan borders."

The main Alberta office is Zone Three in Red Deer; other zones are located in Edmonton, Fort McMurray and Grande Prairie. Collectively, they are known as CASARA Alberta.

Membership is open to aircraft owners and pilots, and those who wish to receive training as navigators or spotters. All are trained in aviation safety, meteorology, survival awareness, and search techniques and procedures. For pilots, the training program requires at least 50 hours on the type of aircraft they are flying. Basically, this means that if a pilot has been flying a Cessna 172 and then buys a 182, they have to put 50 hours on the 182 before they can fly for CASARA. All aircrew must be a minimum age of 18, and sign a medical self-declaration stating they have normal colour vision and 20/20 vision, and are free from any chronic medical conditions that require regular medication.

Further, CASARA pilots must have a valid licence and a current medical certificate, and are required to take spotter and navigation training to Level Three. Following that, they take a test flight in designated training areas with a certified instructor. They learn to follow various search patterns, such as Creeping Line Ahead, Parallel Track, Expanding Square, Sector Search, River Search, Mountain Contour and Track Crawl. This latter manoeuvre would be used, for example, if someone had filed a plan to go from Lethbridge to Calgary and then didn't show up in Calgary. Searchers might start in Calgary, go three miles left of track south from Calgary to Lethbridge, then proceed three miles past, and come back north on the opposite side. This would usually be flown at a fairly high altitude to give large area coverage.

Each aircraft on a CASARA search carries a crew of four: a pilot, a navigator and two spotters. Most volunteers have First Aid qualifications, and each aircraft carries an approved First Aid kit. "We are called 'search and rescue' but we never used to do 'rescue', we used to just do searches," Garry says. "However, a lot of times when you get an ELT – Emergency Locator Transmitter – signal, the transmitter might be out in a field somewhere, and so this call could require a rescue as well." Garry holds a St. John's

Ambulance First Responders First Aid certificate. However, if CASARA is called to what is a good possibility of a crash, STARS [Shock Trauma Air Rescue Society] air ambulance is usually alerted at the same time.

"They've experimented with having us carry rations that can be thrown down to accident victims if they're in an inaccessible crash site," Garry says, "but the military hasn't been in favour of us throwing bundles out of airplanes."

"You can only throw so many pizzas out of an airplane," he laughs. "But I think one of the reasons the military objected to throwing rations out of the airplane is we might get too accurate and hit somebody on the head and that would be a major problem.

"However, we've got the program approved again, and the kit we might throw down is mainly what it takes to survive for a short period of time, such as two-way radios so they can talk to us, and some blankets. It's a question of 'how far can they walk to go and get a First Aid kit?' The military is also working on approving different types of food rations that we might carry."

Flight Services are the first to initiate a report of an aircraft that has missed its ETA. "When you file a flight plan in Alberta, you do it with the Flight Service Station in Edmonton," Garry explains. "If nobody closes that flight plan, they'll first call that airport. If no one answers, then calls go through different phases depending on where the flight plan was scheduled to close. The RCMP might be called and asked to just go over to the airport to see if that aircraft is there. We become involved if the RCMP can't find it. It's generally a pretty good system, but some pilots do forget to close their flight plan and just go home. That can cause a lot of problems.

"The pilot can now get charged for that," Garry adds. "They never used to, but in the past couple of years they're charged about $250 for failing to close their flight plan."

So, when someone is reported missing, the first hour involves a Communications Search. "They'll call blindly on the 126.7, the en route frequency for transiting aircraft, to see if anyone has seen or heard anything," Garry explains. "The airlines, or anyone wishing to monitor 121.5, may do so if they have enough radios and they still report hearing an ELT on 121.5 and that got forwarded to the Joint Rescue Commnication Centre [JRCC] in Trenton, Ontario."*

* As of February 1, 2009 the satellite monitoring of the emergency frequency 121.5 Mh was discontinued, replaced with a newer system monitoring 406 Mh. Non-satellite monitors, both aircraft and ground, continued to monitor 121.5. The Canadian government, with the recommendation of the International Civil Aviation Organization and the insistence of the Department of National Defence, decreed that all Canadian-registered aircraft must be equipped with this new 406 device, some immediately, some up to two years from then. However, this provision was not enforced with U.S. and foreign aircraft transiting Canadian airspace (typically, 63,000 flights annually), and in view of this inconsistency, the government put a moratorium on this change.

"After that hour, if nobody has any ideas on how to contact the missing person or aircraft, then it becomes an Alert. That's when CASARA will probably be called, and/or the military. If there is an ELT signal on that pass, they'll quickly activate a search. Then it will become a scramble to launch aircraft."

Emergency Locator Transmitters transmit an electronic signal that can be detected by satellite or aircraft/ground homers. These fixed or portable mechanisms can be activated manually, or automatically on impact, or when immersed in water.

While Air Traffic Control centres will assist in the communication, all the orders now come from the JRCC.

"The military will give us not necessarily an area but maybe just a location, a latitude and longitude position," Garry explains. "In most cases we then follow the Expanding Square pattern. We start at one mile and expand in a square until that becomes 20 square miles. They'll give us a fairly high search area and will tell us how many miles to go out before it's thought to be enough. I might, for example, receive instructions to go to west of Bragg Creek [30 kilometres southwest of Calgary] and then to fly an Expanding Square pattern at 10,000 feet to give the area maximum coverage."

Because the military C-130 Hercules aircraft covering Alberta searches comes from Winnipeg, and usually from a high altitude, CASARA aircraft are generally the first to reach a search area. The military then contacts CASARA searchers by radio to find out where they're at and what they've done.

As Zone Two Commander, Garry Wutzke is the first to get the calls, and so the responsibility to find a crew also initially rests on his shoulders. "I've tried to give the pager to somebody else, but it seems to be a hard thing to give away. I can give it to someone else for a long weekend if I'm going away, but they are eager to give it back when I'm back home. It's a lot of pressure to get everything organized in a timely fashion. They want us to be launched within an hour and if you're at work it can be quite a challenge."

The number of calls can vary from four or five a month to none at all. "We're not like STARS ambulance where they have shifts and are sitting there waiting for calls, so we can't be on stand-by all the time."

In September of 2008, the Pincher Creek RCMP detachment called Garry, but he wasn't available to go. Like many CASARA volunteers, he has a day job. "I'm a heavy duty mechanic. I had something all apart and the unit I was working on had to go out on a job, so I had to call somebody else to respond. First of all I had to find an airplane, and an available pilot. I called one guy and he didn't call me back for half an hour. I was getting nervous and was considering calling other resources, but he finally called me, and went out and searched for five hours with a full crew that day for a missing person in a vehicle. The RCMP had searched all day on Saturday and Sunday, in a helicopter. We got into the search on the Monday.

"It's very hard to find a person, especially if it's an 'unco-operative target'," Garry adds. "A live person who wants to be found is what we call a 'co-operative target.' When a person is deceased he's not going to wave a blanket, and neither is a person who doesn't want to be found. In this particular case the RCMP had suspected suicide, so that was an issue we were informed of when we were tasked in this case.

"So our people went out to search for this vehicle in the area the RCMP had indicated, mainly the forestry road starting west of Bragg Creek all the way down to Waterton National Park. It took about three hours to cover that area. We stopped in Lethbridge to get more fuel and food, and then headed back to the Porcupine Hills and did contour searching around the road areas until dark. On our last call to the RCMP at 10:00 that night we were told that everything was to stand down because the vehicle and a body had been found off the road in very thick bush area. A landowner had come along and found them."

A similar search request came just after the Pincher Creek call. "On October 10, 2008, we were tasked to search for a hunter who was reported missing near Sundre, but the Red Deer CASARA group looked after that search. In this case, the body of the hunter was discovered; he had been killed and partly eaten by a bear.

"Normally the military will call whoever is in that area, but I talk to controllers at Trenton so often that they seem to call us automatically even when it might be outside our area. I don't usually ask why they don't call another area, I just say, 'Yes, we'll look after it.'"

When calls have come in to one zone commander that should actually be directed to another, the main thing is to get a search underway, quickly.

"When they get a hit from an ELT, the location is not always accurate, especially on the first one or two satellite passes. It can sometimes be off by maybe 50 miles. On December 3, 1993 there was a signal that sounded from near Claresholm, which is in our area, and at that point we were on standby. So we got together up at the Calgary airport, and sat waiting for the second call when Trenton was going to get another satellite pass to identify where it was coming from. By the time the second pass was received, it was identified as being north of Calgary in the Three Hills area. Trenton should have called Red Deer CASARA but because we were already sitting at the airport with a crew, they said, 'Just go.'

"So we went up to Three Hills, circled over the airport, and it was pretty obvious that the ELT signal was coming from there, so we landed. We started calling the numbers that were in the flying club terminal, and it was in a Prairie Bible Institute hangar! We got them out to open the hangar and the first airplane I saw, a Cessna 150, was badly wrecked. The propeller was bent, the wing was bent down, and I thought, 'Holy cow, I think we've found our problem here.' It had skidded off the runway, hit the ice and slid right into the snow. Nobody was injured so they'd just picked it up and hauled it into the hangar. Meanwhile, the ELT was going off!

"The first comment from the fellow was, 'I can't believe that the ELT went off!' and my response was, 'I can't believe that you wouldn't figure that out!'"

Box Canyons

In the afternoon of Sunday, August 28, 1994, Garry received a call about a missing Cessna 150G aircraft (C-FKMJ). The pilot had filed a flight plan from the Airdrie airport to follow the Trans-Canada Highway all the way to Nanaimo on Vancouver Island – a long flight in a Cessna 150G carrying two people, in this case the pilot and his stepson. What was unusual was an ELT had gone off around the Lake Minnewanka area, which many pilots use as a short cut to get to B.C. Garry surmised that instead of following the highway the pilot had probably looked at the map and thought, "This is a lot shorter – we can go straight through by Lake Minnewanka."

The problem was that he didn't go to Lake Minnewanka. He went to the South Ghost Lake instead, and found himself enclosed in a box canyon. While the main fork of the Ghost River enters the eastern shores of Lake Minnewanka at an elevation of 4,850 feet, near Banff, the south fork terminates in a narrow, and deadly, box canyon that is surrounded by 9,000-foot mountain peaks.

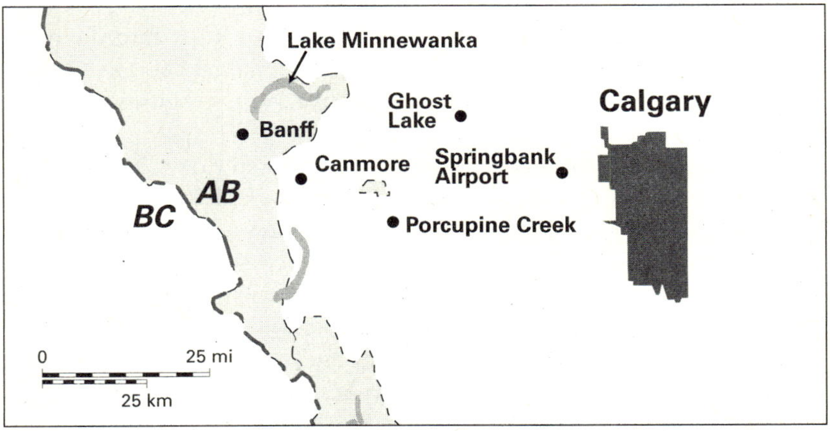

Box Canyon Crashes

"The two routes look very much alike if you're not real familiar with them," Garry says. "When you go into Lake Minnewanka you see water quite soon, and on the South Ghost you can still see water from the river. That box canyon has trapped quite a few people – it got a flight instructor and his family from Springbank a few years prior to this, so I was aware of it."

The CASARA aircraft were first sent to check out the area by Lake Minnewanka, but as they flew by the South Ghost valley they received an ELT signal. "I did not try to go into the valley," Garry says. "Instead, I went up to 11,000 feet and went over top. We came up with a 'lat and long' for this box canyon, and called it in to the Rescue Co-ordination Centre as it was dark now."

The wreckage was found at the south fork of the Ghost River, 10 kilometres north of Canmore, at the 6,300-foot level of the narrow, cliff-edged box canyon. The pilot had obviously been unable to turn around or climb above the canyon walls.

"At that point we feared it would be a fatality, so the Herc was launched and we were sent back home. An hour or so after I got home I heard the Hercules go over, and I thought, 'Oh, this is not good.' They'd dropped flares down into the valley and that didn't work, so at daybreak the helicopters again went out and flew right to location. Both members were deceased."

This search on the South Ghost was one of Garry's first experiences with CASARA that involved fatalities. "When I found out that those people, a 45-year-old man and a 15-year-old boy, had deceased, that bothered me. And the fact that we weren't able to save anybody. That was sad. Our motto is 'So others may live …' but it didn't work. In this particular case, nobody lived.

"Secondly, after the search was over, the military took full credit for the whole thing, even though we'd gone out at night, flying over the mountains, and got them the location of the crash. When it got into the news, CASARA was not even mentioned. I didn't expect them to give us a Purple Heart or anything, but CASARA should have been recognized as being part of this search."

A "somewhat more successful" box canyon search that CASARA participated in took place in the Fisher Range three miles east of Highway 40, near Porcupine Creek in the Kananaskis area. Garry was just sitting down for supper late one Saturday afternoon, May 8, 2004, when the telephone rang. It was CFB Trenton, asking him to get a crew together and prepare for a rescue mission. "I looked outside my window – my best weather-map at the time – and it wasn't a great day for flying. There had been snow showers all day, so when I got the call I was doubtful we could even go."

CFB Trenton told Garry that the pilot, Reg Butler, from Kingston, Ontario, had become concerned that his ELT was not working, and reasoned that before he flew over the mountains in his black and white Piper Cherokee 180 that he should get an appropriate checkout. He stopped in at Springbank Airport where Mitchell Ford, president of Springbank Aero Services Inc., offered to take him for a mountain checkout, which would include a demonstration on how to get out of a box canyon. However, they'd missed their ETA back at Springbank, and an ELT signal had sounded, which automatically prompted CFB Trenton to dispatch a Hercules aircraft from Winnipeg. CASARA was in turn called, along with Kananaskis RCMP, Kananaskis Public Safety Rescue, and Alberta Conservation helicopters, to locate the missing aircraft. STARS air ambulance was also notified, but then they were called away to attend to a gunshot victim in another location, and Alpine Helicopters out of Canmore agreed to stand by.

Flying search aircraft over the mountains at night is always a risk, and flying VFR over the mountains is certainly not recommended. "However, we had the Calgary city lights in view," Garry says, "and because we were there earlier in the evening, before dark, we pretty well knew the altitude of the terrain and had a good picture of the layout of the mountains before darkness actually got there. If it was bad judgement that caused them to get into that situation, we didn't have to add our own to it. Not a good idea."

Through the snow showers, Garry could see a pathway where he thought he'd be able to get into the Kananaskis, so they launched a crew and zigzagged toward their destination.

"When the ELT was picked up we did an X-pattern," Garry says. "We'd centre a needle and go in one direction, east to west and then north to south, maybe not exact because of the angles of the mountains and the valley. We had to come over a ridge and it was very difficult to see. The homer needle told us that it was part way up that ridge, so we knew how far into the valley and what exact ridge it was. There is a cutline in the valley and it was really close to that cutline."

The wreckage was located about 7:30 pm. The mountain flying lesson had gone horribly wrong and the Cherokee 180 had ploughed into trees and rising terrain at the 7,000-foot level of a box canyon.

"We made quite a few passes to get it exact, but we were able to tell precisely where the spot was," Garry says. "Alpine Helicopters went in and found the wreckage right at that certain point. Our instructions from JRCC then were to just keep circling because if there were survivors and they saw an airplane circling it would give them some comfort until the military C-130 Hercules arrived to take over."

Garry circled the area for two hours, waiting for backup to arrive from Winnipeg.

"I guess when you think about it, going out to Porcupine Creek in snow showers sounds pretty dangerous, but it wasn't," Garry reflects. "It was routine, because we didn't go *into* the snow showers to find him, we went *between* them! It was just like this survivor had a guardian angel, because the skies parted right over the crash site. There he was, down in a box canyon. Five miles either way we'd have been in serious snow showers and we wouldn't have been able to go there.

"Alpine Helicopters landed down in the valley and put about a 200-foot sling on the bottom of the helicopter," Garry says. "George Field was one of their people who do high-angle rescues out in the Kananaskis, so he went down and attended to Reg Butler, the owner of the aircraft and the only person who was alive. The instructor was unfortunately deceased from impact with a tree."

Reg Butler, although sustaining serious head injuries, a broken rib and extensive bruising, had climbed out of the airplane and managed to walk a short distance. He said later that he would pass out, come to, and start walking again.

"About that time a C-130 Hercules from Winnipeg arrived to see if there was anything they could do," Garry says, "but everything was under control. I don't know whether I can use this as a 'made my day' incident but we got the job done before the military got there. Even though they'd flown all the way from Winnipeg just to turn around and go home again, they should have been very happy that we got the job done. Because of the location of the wreckage, their SAR-Techs* would have had to jump down to the bottom of the valley and climb about 3,000 feet up the side of a very steep mountain to see what was going on, so I think we did them a helluva favour."

The remains of the aircraft were left intact and two inches of snow fell that night. Accident investigators returned the following day to sift through the wreckage looking for clues to the cause of the accident.

The very grateful survivor expressed extreme sorrow at the death of Mitchell Ford, who he said had taken the controls to fly through the box canyon, commenting "it was a downdraft that got us." Butler could recall the aircraft "hurtling toward the trees." When he regained consciousness he managed to crawl out of the crushed and now wingless aircraft that had come to rest on the ground. Although quite disoriented and in intense pain, he first checked Ford's condition, to find him not breathing and with no pulse. As the fog cleared – both weather-wise and in his own thinking – Butler took the ELT and, although unsure if it was working, tied it as high as he could reach up in a tree. During the flight they had radioed their location to Air Traffic Control so he could only be hopeful that their general vicinity was known and rescue would come. He then found his hiking boots in his luggage, and set off along a creek bed, expecting it would feed into a larger river where he would stand a better chance of being spotted from the air. He found and followed the river for several hours, until he heard aircraft and helicopters overhead. Then he frantically waved his jacket – and was found.

---

* "SAR-Techs" are Canadian Forces specialists who provide on-site medical attention for injured persons in remote areas.

"He was obviously a very capable pilot and knew what he was doing," Butler said of Ford, "[But] these things can happen. It's just tragic, for the poor guy and his friends and family."

The deceased man's brother, Jason Ford, assured Calgary *Herald* reporters that his brother was an experienced pilot who'd made hundreds of trips over the mountains since obtaining his licence five years prior to the accident. "My brother took control of the plane and attempted to pull it up. The plane spun to the side and a tree hit his side of the plane."

For Garry Wutzke and the CASARA crew, it was a bittersweet experience. He deeply regretted that Mitchell Ford could not be saved, but added, "If you save one person, it makes it worthwhile. A lot of times, we make it to the scene two weeks later and discover nobody has made it. The good thing is one person did live."

Box canyons have been the nemesis of many an unwary pilot and have claimed a lot of lives. "We don't go into the box canyons when we're looking for box canyon accidents; we stay up above them and look down. They're just not good places to go."

Garry is thankful to never have been trapped in this way. "We assess the situation from up above before we go down, that's always a good policy," he says. "If you can't get out of it, you don't go into it. There are a couple of life-saving manoeuvring techniques for this situation, but they are extremely risky. Only a very experienced pilot should ever try them, and even then, there is certainly no guarantee of success. If you need one of these manoeuvres, you have violated the laws of mountain flying."

Shortly after the Porcupine Creek incident, Garry was in Grande Prairie to attend an annual SAREX [Search and Rescue Exercise] sponsored by 435 Squadron of Winnipeg. Military detachments from other zones in Canada and the U.S.A. had been invited to take training for mountain searches, and had brought four Hercules aircraft. Garry happened to meet the pilot who had brought the Hercules that night from Winnipeg to Porcupine Creek. "Your CASARA group sure pulled a good one on us," the pilot said in a good-humoured tone, "causing us to fly all the way from Winnipeg on a sightseeing tour."

"You just made my day!" was Garry's response. Finally – military recognition!

Other Common Causes

Risks must sometimes be taken, but CASARA volunteers are sternly advised not to make the same mistakes as the people for whom they're searching. "We stop and survey the situation before we go," Garry says, "so I can't say I've ever had a very frightening experience with CASARA because they've all been well planned, and safety comes first in all situations."

In his work in the search-and-rescue field, he has found the most common cause of aircraft accidents to be weather related. But he adds it is indeed possible that while the pilot is flying in bad weather and icing occurs while in a cloud, something mechanical could also go wrong to stack the cards against a win.

He cites an example of such a combination that felled a centre-thrust twin-engine Cessna T337G Skymaster (C-GXIX) on March 8, 1993. The pilot was CARES/CASARA member Kenneth Bird of Cremona; he was accompanied by a business associate, Kent Ward of Airdrie, whose company, Transfeeders Inc., based in Olds, owned the aircraft.

Ken Bird had registered an IFR flight plan from the Olds-Didsbury airport to Peace River, and was cleared for flight at 11:03 am. Although Air Traffic Control requested that he maintain an altitude of 8,000 feet, a few minutes later it was noted that he was flying at only 6,600 feet. When asked why, the pilot replied that he was "between cloud layers" and would start a slow climb to 8,000 feet. When this wasn't done, the request was made once again. The pilot replied that he was experiencing a rough-running engine but would continue his climb and make a decision on the engine when he reached Rocky Mountain House.

At 11:15 am, when he was reported to be 19 miles from the Rocky Mountain House airport, with bad weather and the engine still running rough, he requested an approach clearance. At 11:27 he asked for, and received, the latest weather from the Rocky Mountain House Unicom. Although there were heavy snow showers and the visibility was recorded as low as one-quarter mile in snow and fog, the pilot stated that he had the ground in sight. Then, C-GXIX simply disappeared.

On learning that the aircraft being piloted by his friend Ken Bird was in trouble, Ruf May, also a CARES member, jumped into Ken's personal aircraft, a Cessna 183 (C-GASS), which had been left at the Olds-Didsbury airport, and flew up to Rocky Mountain House to aid in the search. Brian Moghadam, spokesman for CARES, assured reporters of Ken Bird's competence, adding that just two weeks prior to this incident he and Ken had attended a search co-ordination course in Edmonton.

Garry Wutzke recalls that very little air searching was done to find this aircraft, and CARES was not formally called out. The military came to Rocky Mountain House with the Hercules, but the search was done mostly by foot.

"They had guys tramping the bush around the airport looking for the wreckage, because it was felt that given the shape that airplane was in it wouldn't have gone very far," Garry says. "However, if they didn't find it after the end of the second day, they planned to have a Track Crawl search done all the way up to Peace River because that was their intended destination."

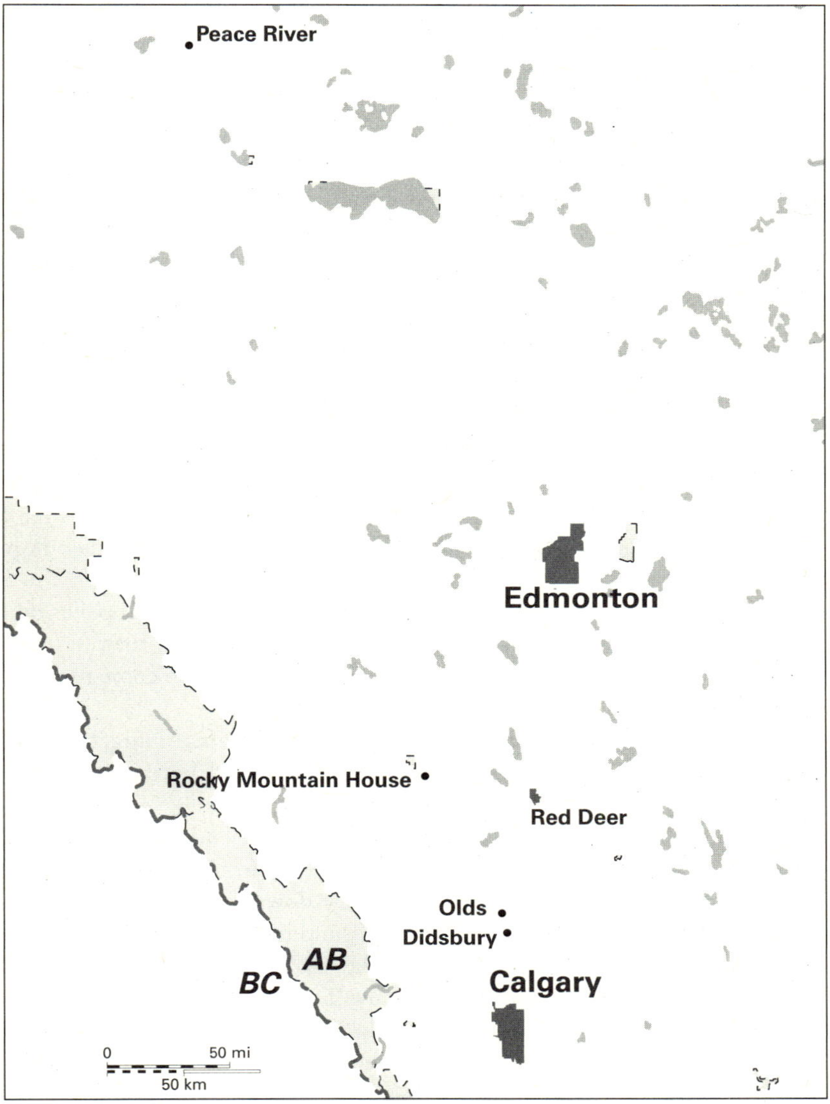

The Ken Bird Accident

On the second day just before dark, and 30 hours into the search, the wreckage was located in the Rocky Mountain House control zone about two miles northeast of the airport. The aircraft had hit a stand of trees on a turn while trying to make an approach into the airport, and then caught on fire. Both occupants were deceased. Because of the severity of the crash, the ELT hadn't activated.

It was later noted that witnesses on the ground had seen the pilot make several passes over the airport, with the front propeller slowly rotating and the aircraft "wallowing" from side to side at slow speed in a nose-high attitude. Indeed, nothing had been going right for Ken Bird and Kent Ward at this crucial time. Another Cessna 337 pilot who came in to the Rocky Mountain House airport at noon said he'd picked up over half an inch of ice on his approach to the airport, and the visibility had dropped to one-half mile in snow showers. "Below Visual Flight Rules" was forecast, with severe icing possibilities in the freezing drizzle, and the Cessna C-GXIX was not equipped for flight into known icing conditions.

Investigators measured the "wreckage trail" as a 10-foot-by-130-foot swath through 75-foot-high trees. Although the aircraft did not break up on impact with the trees, it was completely destroyed by the resulting fire. It was found that the cause of the rough-running engine was a crack in the number four cylinder on the front engine, the result of natural fatigue and overstress. In addition, the front propeller had not been feathered and was on the low-pitch stops at the time of the crash.

Garry Wutzke recalls hearing that "One of the engines was just flopping."

"The propeller wasn't feathered. You have to feather the propeller on a twin engine, or otherwise it's got a lot of drag. Had he done that, he probably could have landed on one engine, but he made a turn to come back and likely stalled in the turn."

In fact, with the front propeller unfeathered and windmilling, the climb rate would have been reduced, a problem that would have been further aggravated by airframe icing.

The CARES community, as well as family and friends, were devastated by this loss. As Brian Moghadam told Calgary *Herald* reporters, "When Bird attempted to bring the crippled aircraft down, he would have had to cope with heavy clouds, snow, strong winds, and the use of only one engine."

Generally, when one of the fellow CASARA members has an accident, the group doesn't get called. "I don't know if that's a written policy," Garry says, "but the military didn't call us to help out to search for Ken Bird. They never specifically said that, but we gathered it by the fact that they abstained from using us."

The reason could be to avoid a repeat of the 1986 triple crashes in the Kananaskis. When the Cessna 182 went missing and friends went searching, only to meet with an accident themselves, the "determination of friends to find friends" could have played a role in this tragedy. High emotions do not fit well with organized programs such as those promoted by the military, and by CASARA.

As with the accident with C-GXIX, the severity of the crash can affect whether the ELT will survive. If it's torn loose from the antenna, that will severely cut down its coverage. "Many times, when it's a crash and burn situation, the ELT succumbs from the fire. They sometimes work for a very short period of time. If the satellite just happens to pick up an exact location, fine; if not, it can go out before anybody gets a bearing on it."

Garry recalls one situation where they found the ELT sitting intact on the ledge of a mountain, although it hadn't worked to alert them to the location of the crash site.

On September 23, 2003, an Alta Flights (Charters) Inc. twin-engine Cessna 414A (C-GVZE) on a cargo and mail run from Cranbrook hit a mountain in the Highwood Range near Longview, about 80 kilometres southwest of Calgary, during a controlled descent to the Calgary airport. Although this was a regular flight, the usual pilot and aircraft (a Piper Seneca) were not available, so the Cessna 414A was substituted. However, the substitute pilot, John Gibbs of Calgary, had flown this route before and was said to be familiar with the Cessna's operation.

The pilot had flown to Cranbrook earlier that day under IFR rules, and in fact a centre-stored IFR flight plan was on file for the Piper Seneca, the usual aircraft. But for the return flight to Calgary the pilot for some reason cancelled this IFR flight plan by telephone with Kamloops Flight Information Centre, and instead filed a VFR plan for the Cessna. Estimated flight time was one hour, at a required altitude of 11,500 feet above sea level.

The last recorded information showed the aircraft flying at 9,000 feet as Gibbs commenced his descent over the Continental Divide, over the high mountain ranges east of the Divide, and into the Highwood Range. Then, with nearby mountains obscured by cloud cover, he hit a west-facing ridge 50 feet below the 8,900-foot top of the mountain. The aircraft completely broke up from the impact, a brief fireball occurred, and items from the cargo, including bank notes, went spilling down the mountainside.

Although the pilot had earlier received Controlled Flight into Terrain [CFIT] avoidance training, the accident investigation could only conclude that he had experienced loss of "situational awareness". This means he became unaware of relationships regarding the aircraft's altitude, the terrain, elevation, or the "lat and long" position.

Many questions came to be asked following this accident, including, "Why would the pilot be flying at just under 9,000 feet when the minimum allowed was 11,500 feet? And why would he switch from an established IFR flight plan to VFR?"

"Nobody could really answer that," Garry says, "except he must have thought that he could do it! This is not the first time an accident like that has occurred. He'd done that mail run flight before, and that can be a hazard when you get too familiar with something. I'm sure he was more careful on the first few flights." It is possible too, that the pilot mistook the Highwood

Valley for the lower altitude eastern slopes west of Calgary, and started a descent over what he thought to be relatively flat terrain.

The military Rescue Co-ordination Centre search and rescue unit were in the air 3.5 hours after the accident was reported. Orin Johnson was one of the CASARA members who accompanied the unit in their Buffalo aircraft, as a spotter. He was making a mid-air "hot swap" with the SAR-Tech, quickly changing positions in the aircraft so there would be no break in the spotting procedure as one person was relieved to go back into the rest area, when the aircraft hit severe turbulence. In a split second Orin found himself forcefully thrown from floor to ceiling inside the aircraft. "That's what it can be like up there in the mountains," he says. "My shoulder ached for weeks after that."

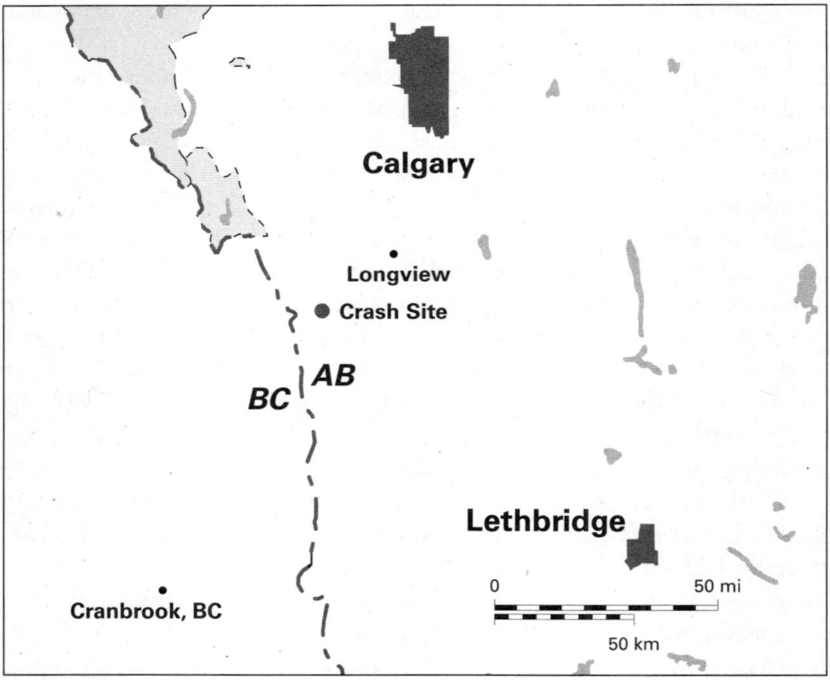

The Alta Flights Crash

The wreckage was discovered on a mountain ridge in the Highwood Range east of the Highwood Valley (between Junction Mountain, seven nautical miles to the northeast, and Raspberry Mountain, 11 nautical miles to the south) some 40 hours after the accident, but it took ground crews more than three hours to climb to the crash site at the top of the steep ridge. Captain Serge Laurin, stationed at the temporary command centre at the Calgary airport, explained that while every attempt is made to get the rescue crew as close as possible to a crash site, because of the rugged terrain in this case the Canadian Forces Griffin helicopter had to set the rescue operation personnel some 1,500 feet below the site.

"Most of the airplane parts went over the other side," Garry says. "The biggest visual item was that orange ELT sitting on the ledge of the mountain, and a burn spot on the mountainside that was being investigated when they found the ELT."

The 31-year-old pilot, John Gibbs, had begun his love of flying with Air Cadets and had won a scholarship to Mount Royal College's aviation program, graduating in 1994. So many questions, so few answers. Just scattered wreckage on a high rugged mountain ridge, broken-hearted family and friends, and puzzled and sorrowful search and rescue personnel who could do nothing but try to pick up the pieces.

Specs for Spotters

To be trained as a spotter for CASARA, one must be at least 18 years old and have 20/20 vision, which can be corrected with glasses. "Unfortunately a lot of our members are seniors, of an age where they need glasses," Garry laughs, "so I'm certainly glad we don't have that restriction."

Spotters are trained in a three-hour classroom session, shown a survival film and spotting techniques, and then taken on a flight. There are two flights of a minimum of one hour each, and a third flight where the candidate is evaluated to discover if they can spot things they should be seeing. A target is planted, clearly identified with bright orange letters, and the candidate is expected to spot it.

Secondly, the spotter must show an ability to call an aircraft around when they see an object, and provide geographic references so the crew can also see and identify it. Their call might be "Turn right … by that grove of trees … by that open field … by that big rock" or "by that elevator, that granary out in the middle of the field." Directions such as north, south, east or west are not used because when your head is down it can be difficult to tell what direction you're going – a fix based on features of the landscape works better.

"We're pretty lenient," Garry says, "but we like to find people who can spot items of interest, which may be pieces of the wreckage we're looking for. You don't want to go past wreckage and miss it. So from that point on, the spotter needs to take two flights per year of not less than an hour each, plus three hours of classroom training. It's not really that tough to maintain their spotter currency."

Passing the spotter academic test, and spotter certification flight and practical test, completes one's acceptance into the program.

The tendency for motion sickness also must be considered when evaluating the qualifications of a spotter. "In that case, you have to put restrictions on that person. It happens once in a while, but in a lot of cases they outgrow it if they're up in the air a lot. When you start navigating or spotting you have to look down at the maps and onto the ground, and that causes some people to get motion sick.

"If that happens, we say, 'Just look out.' However, you have to be careful not to look out when you're in the process of spotting. What you do there is break off the search pattern and give your eyes a rest, so everybody can take a look around. When we search for, say three or four hours, at the end of the legs outside the search area we put the spotters off task – and that's generally for about two minutes on each leg – so they can rest their eyes. The pilot and navigator are still on task, of course, but the spotters are off."

There are not necessarily set teams of the same pilot, navigator and spotters. "We mix," Garry says, "and when we're training it's whoever comes out and needs training in that particular area. Also, some aircraft can carry only a given amount of weight and so we have to divide these four people up into a crew that is within the weight that can be carried. I have a Cessna 182 so I can usually take the big ones."

Garry's aircraft is also equipped with long range tanks and has a fairly large fuel capacity, so under normal circumstances he could stay aloft for close to six hours.

"We try not to go that long, but sometimes if we're over a wreck we have to be able to do our endurance time. I have a fuel flow meter, which is a great asset. I can go up to high altitude and set the speed right down, and I'm good for another four hours before they have to bring out another airplane to spell me off."

He explains that when the replacement aircraft come into a site they always fly in at a different altitude. "If I'm at 8,000 feet, someone will come in at maybe 8,500 – giving at least 500 feet vertical separation. Then I'll give them instructions on what we're looking for, what we're circling, and then I'm off task and they're on task."

The mixed male and female crew, if they're aloft for a number of hours, must eat and drink, of course, and sooner or later go to the bathroom. This latter duty involves a reluctant, but discrete, use of "Little John".

The Vagaries of Crashes

What do searchers look for when an aircraft goes down? What telltale signs, aside from ELT signals, will guide searchers to a wreck site?

Zone Two takes CASARA searchers over an incredible variety of terrain from farm fields to mountain peaks, over lakes and rivers and dense forests.

"If an aircraft has hit the side of a mountain there is often a flash fire," Garry says. "You might spot the signs of a small forest fire, but that's often hard to see. A lot of them go out quickly so there's just a big puff of smoke, but someone on the ground might have seen it. I can think of several accidents where that's all we had to go on: someone had seen a puff of smoke."

The unfortunate case in October 1988 involving 39-year-old pilot Gordon Lloyd Wilder is often referred to by searchers when discussing this type of sign that can lead to the discovery of a crash site.

Wilder, whose family owned the Fairmont Hot Springs Resorts, loved the mountains. After getting his pilot's licence in his early 20s he purchased and flew a number of different types, including an aerobatic Citabria. He often took tourists on glacier sightseeing tours in a ski-equipped Cessna 180 that belonged to his own company, Fairmont Air Ltd. Even in the summertime he would land with skis on top of a mountain in the Kananaskis area so people could get out and walk around and enjoy the view from this unique vantage point.

Wilder's fatal accident did not occur on one of these jobs, however, or during one of his many daring ventures that included hang-gliding, scuba diving, hiking and skiing – or on his ventures to mine the Yukon's gold and fashion his findings into jewellery for another of his companies, Fairmont Gold Ltd.

On Friday, October 21, 1988, a particularly windy day, Wilder flew his girlfriend, an airline flight attendant, to Calgary to report for work on a scheduled flight to London. He then filed a pilot report [PIREP] with the Springbank Airport tower at 1:45 pm, describing the turbulent weather conditions he'd encountered in his one-hour flight from Fairmont to Calgary. Such information is usually relayed by radio to the nearest ground station. The message is encoded and relayed to other weather offices and air traffic service units.

And then Gordon Wilder turned around and flew right back through the turbulence.

A search was hampered by the very conditions that had caused him to miss his ETA: extremely turbulent winds and light snow, aggravated by darkness. The search, when it got into high gear, involved about 60 civilian and 45 military personnel using four military helicopters and two Twin Otters from Edmonton, four civilian search planes and one civilian air communications plane – searching in terrible mountain weather conditions for a little white aircraft, or what might be left of one.

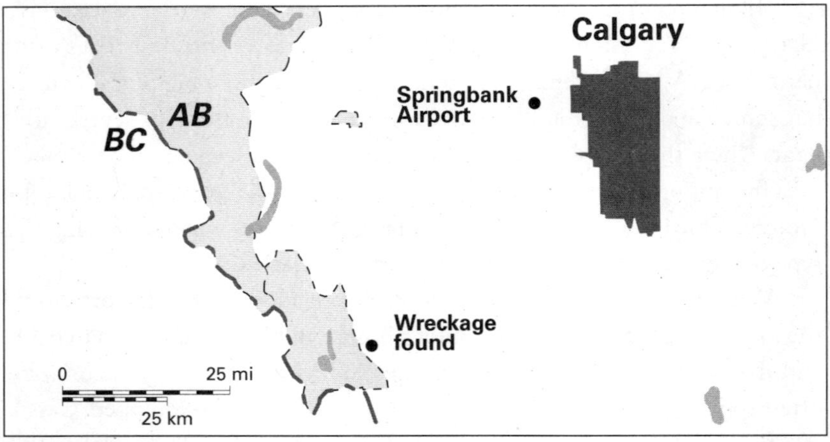

The Gordon Lloyd Wilder Crash

After a day of futile searching, park rangers set up a roadblock in the Kananaskis Country and asked motorists if they'd seen anything. One person said he had seen something odd but didn't know what it was. "I saw a puff of smoke on that ridge right over there," he said, and pointed to a spot high on a mountain. And then two people on a horse trail-riding trip reported they had seen a flash of fire.

A helicopter was sent in with forestry spotters, but after going over the area no less than six times and not finding the crash site, they started a foot search. First of all, someone noted a broken tree. And then, as the searchers tramped along, they noticed that their clothes were getting black from soot rubbing off the brush and trees, evidence of what could have been a flash fire. At 6:30 pm on Sunday, October 23, they came across the still-smouldering wreckage of the Cessna 180 at the southeast base of Elpoca Mountain in the Opal Mountain Range just outside Peter Lougheed Provincial Park, about 80 kilometres southwest of Calgary. The cockpit, fuselage and wings were charred, leaving only the tail section that might have been spotted from the air.

It is believed that when Wilder reached the Elbow Valley in the Kananaskis, an area noted for its high winds, his forward speed would have been severely reduced. Then, when it became so turbulent that he decided to turn around, his forward speed would have increased like a jet. Investigators from the Canadian Aviation Safety Board confirmed that the aircraft was in a 10-degree bank turn when it struck some treetops on the side of a ridge. The aircraft came apart, the fuel tanks were breached, and a fire followed the impact. Evidence showed that the pilot had survived the crash for an undetermined period of time, but perished from his burns before discovery.

"The pilot had walked a few feet out from the airplane, believe it or not," Garry says, "but he was probably in shock because it was a very bad crash. The aircraft was heavily damaged and no ELT was detected; it must have succumbed from the fire."

The puzzle remains why the pilot would register a PIREP and then go right back onto that route again. "Very bad judgement," Garry concludes.

"The same reason, I guess, why people might look at the weather forecast and think, first, 'That's too bad to fly in,' and then their next thought is, 'Well, I'll just go out the airport and have a look.' A chain of events can then occur. They're at the airport, and the next thing they do is pull the airplane out of the hangar. Another link in the chain. 'We might as well put our suitcases in the airplane, because any minute now the weather's going to get better.' Okay. Another link. 'Now let's pull over to the gas pump, make sure we have enough gas so everything will be ready to go when the weather clears.' And it just keeps going, one link at a time, doing things that lead to bad judgement. Next thing you know, they're warming up the engine. Next thing, they're taxiing down the runway."

He recalls a speaker he heard at an air show at Oshkosh, Wisconsin, who was now working for the Federal Aviation Administration [FAA] in the United States. She had been a bush pilot in Alaska, and a number of her friends had been killed in aircraft accidents. She then related some interesting stories that Garry found to be so true, citing scenarios that often resulted in accidents. One was reckoning that "If the weather's too bad to go IFR, I'll go VFR." It should, of course, go the other way around – "If it's too bad to go VFR, I'll go IFR – or not at all." But these pilots figure that if the weather is below minimums for flying IFR and conditions for VFR aren't as strongly restricted, they can go.

The speaker then showed a picture of an airport with airplanes parked all around, and the fog down to the height of the airplanes. And then she said, "You'll notice on one airplane the propellers are turning! They were getting ready to go – and they did."

Garry explains that when you file for an IFR flight plan, stricter regulations must be followed and Flight Services simply won't allow someone to file an IFR flight plan if the weather is below minimums. "You aren't supposed to file a VFR flight plan under minimums either," he says, "but some just do it. It's up to the pilot's discretion. The Flight Services Station [FSS] might say, 'VFR is not recommended,' but they can't stop you from going, whereas with IFR, they could. They just won't file the plan.

"However, since October 2005, FSS no longer uses the term 'VFR not recommended'. Now, the decision rests with the pilot. The FSS gives a weather briefing and if ceilings are 400 feet, for example, they may suggest waiting a while but if you decide to go despite the forecast that is up to you."

The Graveyard Spiral

CASARA volunteers are all too familiar with what is called the "Graveyard Spiral" that can occur if a pilot gets into a cloud where there's zero visibility.

"In these cases, there are a lot of books that say 'You have 144 seconds to live' if you don't take remedial action immediately. That is the estimated length of time it takes for a VFR-rated pilot to become disoriented, from the time this happens to the time he hits the ground," Garry says. "However, there are exceptions to every rule. Some pilots are able to maintain their situational awareness and pull through it, but for others, they don't realize they're in a Graveyard Spiral until it's too late. They think that what they're doing feels right, like normal flying, until all of a sudden the ground comes up, or they happen to look at the altimeter and see it unwinding – but by that time it's too late because their speed is so great that pieces of the airplane start coming apart."

So, the pilot is flying along in cloud and he can't see anything. He quickly loses spatial orientation, doesn't know if he's up, on a slant, or going down. Why doesn't he check the instruments?

"They've obviously taken themselves off instruments," Garry surmises. "Some might think their instruments are off kilter. Maybe he's had trouble with this instrument, and for whatever reason he thinks it isn't reliable anymore – because what the instrument says just doesn't feel right. He doesn't believe the instruments, and that's where things go wrong. The aircraft slowly starts banking, goes into a turn and then starts dropping, faster and faster. When it comes out from underneath the cloud and the pilot realizes what's happening, the aircraft is usually going so fast that structural damage occurs."

Garry explains that every aircraft is rated for a certain speed, termed the VNE or "Velocity Never Exceed". There are a lot of V speeds in aircraft: a VFE for Flap Extension, or in aircraft with retractable gear there's VGE for Gear Extension. Test pilots have discovered the various restrictions, and warned: "Don't exceed that speed because things will start coming apart." Often, VNE is based on factors like the design of the aircraft.

"There are some aircraft now that if you enter the cloud you can take your hands off the control column and steer with your feet. Watch the directional gyro and keep it on the heading, and it won't let you run the risk of steering into a spiral," Garry says. "So if a pilot were to stay on instruments and do what they say, especially in the turn cycle, some of these tragic accidents could be avoided."

Basically, in a Graveyard Spiral, the aircraft will descend faster and faster as it spirals down. When it hits the ground, some heavy parts such as the engine could be buried deep in the earth. Search spotters might note a sign as simple as a broken tree, or an extraordinary gathering of animals or birds. They might see a glint of light between the trees from some small airplane parts that the sun happens to hit.

"A spotter has to recognize any and all these things and check them out. It could be our only clue to finding the wreckage within a reasonable time, to help any survivors that may be down there."

Another example of an unsuccessful search is a case referred to as "SAR [Search & Rescue] Henderson". On March 19, 1992, pilot Delmar Henderson and passenger Hugh Bothner were in Henderson's single engine Beech Bonanza B35 en route from Rainbow Lake, 400 miles north of Edmonton, to Springbank Airport west of Calgary. They had been working in the sour gas industry, on contract for Dedicated Technology.

At 2:30 pm the pilot made radio contact with the Grande Prairie flight area, stating they were at 13,000 feet in the proximity of the Smoky Tower. Then, they disappeared off radar.

The Grande Cache area they would be flying into was experiencing weather considered poor for flying, with cloud and snow. When they didn't come in to Edson to refuel, over 40 military and civilian search aircraft were organized to scour the surrounding foothills and mountain country, to no avail. By Friday, March 20, 11 private CASARA aircraft from Edmonton plus two military Hercules and three Twin Otters went out, with five more from Calgary CASARA joining them the next day. Brian Moghadam, spokesman for Calgary CASARA, described their initial requested search area as being from Innisfail south, west along the mountains to Ghost Lake, and east to Calgary. Later, they were assigned to search the Rocky Mountain House and central Alberta area, while the military and Edmonton and Grande Prairie CASARA groups covered the northern areas. When nothing was spotted, all searches – now comprised of several military and some 30 civilian aircraft – concentrated fully on the Edson to Grande Prairie sector. A ground search was considered, but the rough country in which the aircraft might have disappeared would make that a tough assignment.

"Normally you'd think there should be some trajectory where the radar track stopped, but that search went for two weeks," Garry says. "They used infra-red and found every moose in the country, but because the two people in the aircraft – Delmar Henderson and Hugh Bothner - had deceased, it didn't record any warm bodies. Two years later the wreckage was found by some hunters, right where it was supposed to be. From what we could gather it was a straight and normal flight, until he got into snow showers and encountered severe turbulence that did structural damage to the airplane."

Some disappearances have never been found or resolved, such as that of Gavin Edkins, who in 1996 was flying a Cessna 150D from Fort Smith. Edkins stopped to refuel in Fort McMurray and then disappeared en route to Red Deer.*

---

* This case was featured in "The North Reclaims its Own", pp. 69-90, in *Lost: True Stories of Canadian Aviation Tragedies,* by Shirlee Smith Matheson.

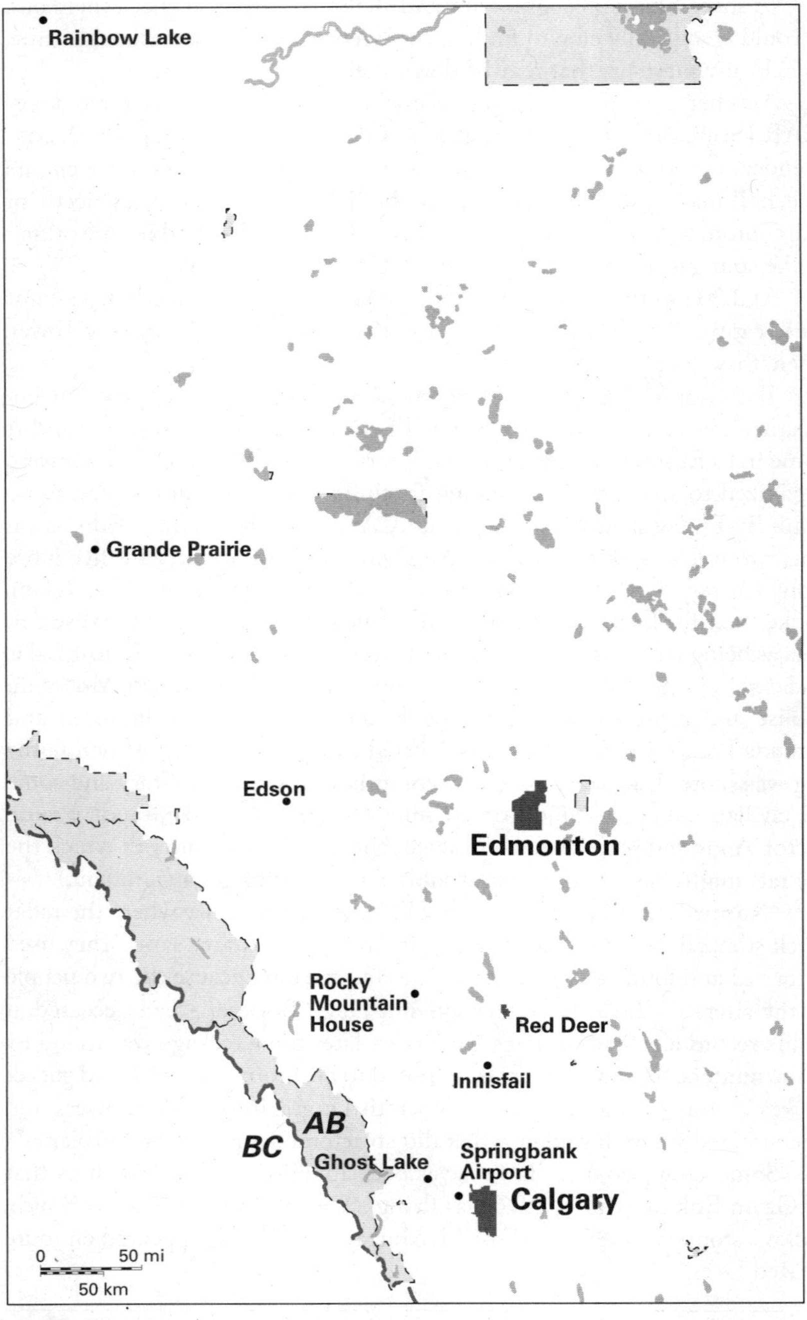

The Delmar Henderson Crash

Garry notes "There have been a lot of clues on that one. We searched for just a little better than a week, and then the search went into different phases. Edmonton CASARA re-enacts the search every year as a training exercise because they are still frustrated and saddened that the disappearance has never been solved. They go into what they feel is the hot area, in this case around Lac La Biche, and re-do the search. But nothing has come up."

In British Columbia something like 110 airplanes are recorded as being lost and never found. "There is probably near that number in Alberta too, some going back for years. Over my 20-some years with CASARA, there have been so many."

The Good Ones

One call, where CASARA search crews didn't have to even get off the ground, was an alert on September 17, 2008. A Beech Baron had taken off on a night flight from Medicine Hat Airport en route to Fort St. John, B.C., with five people on board when it met with an accident.

"They'd just departed Medicine Hat when the pilot called in a May Day that they were going down, probably because the engines quit," Garry says. "We were called to respond, along with STARS air ambulance and the military from 435 Squadron in Winnipeg – whoever got there first. We were already in our airplanes ready to go, complete with equipment such as night vision goggles as it was about 10:00 on a nice moonlit night.

"Then we were called off short when the Rescue Co-ordination Centre in Trenton advised us that the aircraft was found – right in the South Saskatchewan River! Two people were able to walk out of the wreckage and identify themselves as those involved in the crash. The joy we felt when we learned that everyone was going to be okay with the medical help available! STARS ambulance attended, and one person was airlifted to hospital. There were three other people injured but not severely, and nobody died."

A month later, on October 15, 2008, Garry Wutzke received an email from Mallory Smith, one of the persons involved, thanking everyone who had been alerted to their plight and had been ready to go, or was directly involved in their rescue:

> Garry,
>
> I am not sure there is an adequate way to say thank you to all of those, like yourself, who responded in such a rapid and professional manner during my and my friends' time of need. Since I was a young teenager and subscribed to my favorite aviation magazine, Flying, I have only read of experiences like the one I endured the 17th of September!

Our group had been in Bow Island that day, invited to sing for an acquaintance's funeral. We hurried back to Medicine Hat to start our journey back home to Fort St. John, B.C. We pre-flighted our Baron and headed out to the advised runway 09 (we requested 03 but agreed to depart to the east). We had no sooner lifted off and raised the gear when we felt that dreaded yaw accompanied by the unmistakable sound of our right engine surging. My friend, the pilot, instinctively switched on the fuel boost pump to the engine and trimmed the aircraft to compensate for the loss of power on our right side.

As I am sure you know, the worst phase of flight to lose an engine was where we were at that time! We desperately needed airspeed in order to climb and needed altitude in order to drop our nose a little to get that airspeed! We had neither that night.

If you Google Earth the airport (YXH) directly off 09 you will see a small ravine that runs north and south. We gently turned to the north in this ravine and headed down towards the river in a desperate hope we might find some room to get some airspeed, and also perhaps in the cooler river air get a little better performance out of the sole operating engine. We made one short attempt to restart the failed engine, to no avail. We then decided to feather the engine and give full attention to flying the aircraft that was just on the verge of VMC [Visual Meteorological Conditions].

One of the remarkable aspects of the entire experience was that during all this time (from take off to our ignoble landing in the South Saskatchewan River we were airborne for only seven minutes!) there was no fear or visible anxiety displayed by any one of the five of us on board at anytime!

Once the pilot and I realized that we were not going to get the aircraft to climb, we decided to use the river as our landing site. We hoped to skip on the water to a large gravel/sand bar on the inside curve of the river. There was a small cluster of cottonwood trees on the edge of the river that we had every intention of going over as opposed to through! Once we hit the trees, though, gravity was the predominant force in the remainder of our flight!

We hit the water hard and flat in a straight and level attitude. The force of the impact opened the rear cabin doors, making exiting the aircraft easy for the three individuals in the rear of the airplane. I, on the other hand had been knocked unconscious upon impact and needed to be removed from the aircraft by the others. The pilot and I received significant facial and scalp lacerations but otherwise were relatively unharmed. The two aft-facing occupants were without injury and the 5th, a front-facing fella, incurred some ankle and foot injuries as a result of bracing for impact.

The water level of the river was the lowest it has been in years. The aircraft came to rest in the middle of the river and we found the depth to be approximately a meter deep! There was a fast current but all of us made it to the shore with little problem. It was on the bank of the river where three of us stayed while two others walked out to find help. Those were the most distressing three hours I have ever had to endure! But together we joked and shivered until help began to arrive at 10:00 pm that night. Four of us were taken to Medicine Hat Regional Hospital that night, and the individual with the ankle injuries was airlifted by STARS to Calgary. The staff at the hospital was so kind and compassionate; we were humbled by their genuine concern and care.

All of us are home now and healing gradually. The return to "The Hat" was very special to us as we wanted to have the opportunity to acknowledge the efforts of so many in our rescue and treatment. Please extend our deep appreciation to those of your organization who, though doing their job, were heroes in our eyes that night. I thank you, my family thanks you.

Sincerely,
Mallory Smith

"It was a very good sort of survival, at night, down in the river, and everybody lived," Garry says. "It's one of the more unusual stories, as the crash had the potential to be very serious. It was a 911 alert. When we got the call at 9:30 that night, they said, "GO!" But it had the potential of a search and a rescue."

The other side of the coin is when CASARA and the military are certain that all avenues of a search have been covered, the area has been exhausted, and it's known that they are no longer looking for live bodies. The case is then turned over to the RCMP as "missing persons". Once the military stands down, CASARA stands down as well. The only exception is if the RCMP tasks CASARA to continue under a Memorandum of Understanding that the organization holds with the RCMP regarding such searches.

It Could Have Been Me ...

On November 8, 2008 an accident occurred that shook the central Alberta community, including CASARA members, when one of their own met with a fatal accident while returning from a CASARA night flying training exercise.

Herluf Neilsen, a 67-year-old "flying farmer" from Spruce View just west of Innisfail, Alberta, loved aviation. He was one of the founding members of the Innisfail Flying Club, he promoted the Young Eagles program that gave first flights to over 1,000 Grade Six students, and was an active member of CASARA Red Deer Zone Three.

Coincidentally, the December 2008 issue of COPA *Flight*, the publication of the Canadian Owners and Pilots Association, carried an article earlier written by Kristine Neilsen. "Chronicles of a Cessna" detailed the trip of a lifetime taken by Kristine and her father Herluf in 2007, from their home in central Alberta up to the Yukon and Alaska. "This is the life, dear Kristine!" Herluf often said, and it was.

That fatal night of November 8, fog moved in quickly as Nielsen was transiting home to the airstrip and hangar on his farm at Spruce View in his Cessna 182 following the CASARA exercise at Red Deer. He called his wife, Alice, to say that because intense fog had suddenly rolled in, he'd decided to land at the Innisfail airport instead. When he became overdue Alice called the airport to report him missing.

At 4:50 am on November 9, Garry Wutzke received a call from Joint Rescue Co-ordination Centre at CFB Trenton, who had been notified of the situation at midnight. A C-130 Herc had been launched out of Winnipeg and would land in Calgary to pick up spotters at 7:00 am. Garry planned to wait an hour before calling out members since he had a two-hour window and the weather was extremely poor with fog and low visibility. Then he was informed that a ground search had been launched. At 5:20 am the wreckage was found in a farmer's field just south of the Innisfail airport, but no body. At 5:40 another call from JRCC Trenton indicated that the body had been located near the wreckage. The succinct terminology told the whole story: "Code Black".

The phrase "It could have been me" has been uttered numerous times as airborne humans miraculously have managed to pull their aircraft out of a deadly Graveyard Spiral, or escaped weather that seemed destined to prematurely terminate their flight. And so many aviators feel it is indeed their duty to use their talents and time, and money, to become members of organizations such as CASARA in order to promote aviation safety and provide search – and hopefully rescue – operations to help those who have met with misfortunes.

Their efforts sometimes result in sorrowful learning experiences, and at other times rewards beyond description. It is somewhat ironical that the ultimate measure of success for the organization would be that flyers would never have to deal with them at all – in fact, Patrick Fahy, CASARA-Alberta president, often gives a talk entitled "How to Avoid CASARA" emphasizing the importance of aviation safety.

For people like Garry Wutzke and other CASARA volunteers who divide their time between their home and day-job commitments, spend time and resources on the upkeep of licences and machinery, and combine their joy of flying with a necessary and benevolent volunteer activity, such efforts can be stressful yet life-enhancing, in more ways than one.

Garry hasn't kept many photos of search and rescue missions. "We're more picture-conscious now, and sometimes I wish I'd been doing that back in early years," he says. Then he closes a log book that is filled with records of search flights that brought both joy and sorrow: accidents where survivors desperately clung to the hope of being found, and were; and others that had forever ended someone's dreams.

"All I can do is tell you stories," Garry says. "And keep on flying … so others may live."

Aviation Abbreviations and Acronyms

AD....................... Airworthiness Directive
ADF..................... Automatic Direction Finder
AES...................... Atmospheric Environment Service
AFB...................... Air Force Base
AME.................... Aircraft Maintenance Engineer
ASL...................... Above Sea Level
ATC...................... Air Transport Committee; also Air Traffic Control
BCATP................. British Commonwealth Air Training Plan
C. of G. .... Centre of Gravity
CAPA................... Canadian Airline Pilots Association
CARES................. Civil Air Rescue Emergency Services [this organization superseded by CASARA]
CASARA.............. Civil Air Search and Rescue Association
CFB...................... Canadian Forces Base
CFIT..................... Controlled Flight into Terrain
COPA................... Canadian Owners and Pilots Association
CP (also CPAL)..... Canadian Pacific Air Lines, Limited (confirmed as official form in CP Annual Report)
DART................... Deep Access Reconnaissance Television
DEW Line............ Distant Early Warning Line
DFC..................... Distinguished Flying Cross
DI .... Daily Inspection
DOT..................... Department of Transport
ELT....................... Emergency Location Transmitter
ETA...................... Estimated Time of Arrival
FAA..................... Federal Aviation Administration (United States)
FSS...................... Flight Services Station
GPS..................... Global Positioning System
HP....................... Horsepower
ICAO................... International Civil Aviation Organization
IFR....................... Instrument Flight Rules
INS....................... Inertial Navigation System
JRCC.................... Joint Rescue Co-ordination Centre
LORAN................ Long Range Navigation
LST...................... Landing Ship Tank
MAN.................... Missing Aircraft Notification
MATS.................. Military Air Transport Service (division of USAF)
Medevac............. Medical Evacuation
MPH.................... Miles per Hour
MRCC.................. Military Rescue Co-ordination Centre
NDB..................... Non-Directional Beacon
NORAD............... North American Aerospace Defense Command

NOTAM .... Notice to Airmen
PIREP .... Pilot Report
PPC....................... Pilot Proficiency Check
PPCLI .... Princess Patricia's Canadian Light Infantry
PWA .... Pacific Western Airlines
QCA .... Queen Charlotte Airlines Limited
RCC..................... Rescue Co-ordination Centre
RAF...................... Royal Air Force
RCAF .... Royal Canadian Air Force
RCMP .... Royal Canadian Mounted Police
RPM..................... Revolutions per Minute
SAIT..................... Southern Alberta Institute of Technology
SAR .... Search & Rescue
SAREX .... Search and Rescue Exercise
Sea-Tac .... Seattle-Tacoma Airport
SOP .... Standard Operating Procedure
SOS .... Save our Souls [also Save our Ship]
STARS.................. Shock Trauma Air Rescue Society
TCA .... Trans-Canada Air Lines
UHF .... Ultra High Frequency
USAF .... United States Air Force
VFE...................... Velocity Flap Extension
VFR...................... Visual Flight Rules
VGE..................... Velocity Gear Extension
VHF .... Very High Frequency
VMC.................... Visual Meteorological Conditions
VNE .... Velocity Never Exceed
VOR..................... Very High Frequency Omni Range
WFL..................... Wetter-Funkgerät-Land [Weather Transmitter – Land]
WFS..................... Wetter-Funkgerät-See [Weather Transmitter – Sea]

REFERENCE SOURCES

Deadmen's Valley

McLeod Brothers, Mad Trapper

Books

- North, Dick: *The Mad Trapper of Rat River*, Macmillan of Canada, Toronto, 1972.
- Patterson, RM: *Dangerous River,* George Allen & Unwin Ltd., London, 1954, pp.21-28.
- Turner, Dick: *Nahanni*, Hancock House Publishers, Surrey, B.C., 1975.
- ---- *Wings of the North,* Hancock House Publishers, Surrey, B.C., 1976.

Newspaper

- Calgary *Herald*, June 24, 2006, p. A12: "Jean's plane used in search for body".

Periodicals

- *New Trail* magazine, University of Alberta; Autumn 2008; "Mad About You", Dr. Owen Beattie.
- *RCMP Quarterly*, January 1960; "Valley of No Return", RCMP Cst. T.E.G Shaw.
- *RCMP Quarterly*, January 1961; "South Nahanni", RCMP Cst. T.E.G Shaw.

Reports

- Neary, Sergeant Major (later Lt. Col.) I. "Nash", Warrant Officer in charge of the Royal Canadian Signals Radio Station, Aklavik, NT: Report dated February 21, 1932, to Director of Signals, Elgin Building, Ottawa re: Mad Trapper events, capture and aftermath.
- Neary, Lt. Col. I: extract from October 1948 "Notes of Interest" (Northwest Territories and Yukon Radio System publication) issued by the Command Signal Officer, Western Command Headquarters, Edmonton, and prepared for exhibition in the Communications and Electronics Museum.

Five Prospectors and 'Starvation Cabin'

Interviews, correspondence

- Jim McAvoy, personal interviews July 19, 2003 and other dates.

Periodicals

- *North*, Vol. VIII, No. 3, May-June 1961, "Valley of Lost Reputations", RCMP Cst. T.E.G Shaw.
- *RCMP Quarterly*, January 1961; "South Nahanni", RCMP Cst. T.E.G Shaw.
- *Star Weekly,* October 8, 1960, "The Jinx of Headless Valley Kills Again".
- *Up Here* magazine, May/June 2003, "The Jinx Strikes Again", Randy Freeman.

Other sources

- Court documents.

Blake MacKenzie

Interviews, correspondence

- Steve Villers (Villers Air Services), emails from January to April 2003.

Newspaper

- Edmonton *Journal*, January 23-24, 1962.

Periodical

- *Weekend Magazine*, No. 38, 1963, "Diary of a Mystery", Edward Romaine.

Other sources

- Department of Transport Air Services, Civil Aviation Branch, Accident Report re: Cessna 180, CF-FYN.

Gregson's Gold

Books

- Jackson, Susan (editor): *Yellowknife, NWT: An Illustrated History*, Vol. I Yellowknife History Series, Nor'West Publishing, 1990, p. ?13.
- Miller, Dave: "Gold Heist" (Canadian Broadcasting Corporation audio production featuring comments from Max Ward, John Parker, Shorty Brown and Doris Gregson), January 2002.
- Unattributed: *The Story of Discovery,* Women's Institute, Yellowknife, NT.
- Ward, Max, *The Max Ward Story,* McClelland & Stewart, Toronto, 1991, pp.128-131.

Newspapers

- *News of the North,* Vol. 10, No. 9, page 1, July 9, 1954, "$50,000 worth of gold spirited away", and page 6, editorial, "And of course – gold bricks".
- *News of the North,* Vol. 13, No. 4, June 27, 1957, "Tony Gregson, Without Bricks, Picked Up in Australian Gaol".
- *News of the North,* August 22, 1957, "Gregson's plea is guilty of theft".

Periodicals

- *Up Here,* July/August 2001 (Vol. 17, No. 5): "The Great Gold Theft", Randy Freeman, pp. 54-56.
- *Up Here,* January/February 2002: "Speaking Up –The Great Gold Theft", comments by David Geddes.
- *Up Here,* May/June 202 "Speaking Up – The Great Gold Theft", comments by Ed Overbo.

The Unfair Weather of Mount Fairweather

Book

- Milberry, Larry: *The Canadair North Star*, CANAV Books, Toronto, 1982, p. 162.

Other sources

- Corcelli Research (John Corcelli, President), Toronto.
- Department of Transport records of investigation into the disappearance of the CPAL DC-4, (CF-CPC).
- Carl Mills of Banshee Publications, North York, ON.
- Dan Ryan, Canadian aviation artist, Coquitlam, B.C.
- Jerrold E. Vernon, researcher, historian and president, Vancouver Chapter, Canadian Aviation Historical Society.

The Book of Broken Dreams

Dying for Love

Newspapers

- Canadian Press article about Ernie Simmons of Tillsonburg, ON published in the Vancouver *Province* and other CP publications, August 29, 1970; in the Toronto *Star* (newspaper) September 4, 1970; and in *Canadian General Aviation News*, July 1986, page 53.

Whether the Weather

Report

- Melissen, Melis: transcript of magnetic log at Campbell River Flight Service Station September 2, 1983; interviews and correspondence with Bill Cove; other reports and communications.

The Vicious Vortex

Report, other sources

- "Aircraft Accident Report" for Johnny Boak; interviews and conversations with Bill Cove; other reports and communications.

The Loggers' Discovery

Interviews, correspondence

- Joe Van Es, telephone interviews.

Periodical

- *BC Aviator,* January/February 1994, "A Mile in Rain and Fog", Jack Schofield.

The 37-Mile Flight

Book

- Matheson, Shirlee Smith: "The Paper Bag Prince" in *Flying the Frontiers Vol. III – Aviation Adventures Around the World*; Detselig Enterprises/Temeron Books, Calgary, 1999.

Interviews, correspondence

- Joe Mugridge, correspondence, February 4, 16, 27; March 16, 25; April 7; November 29, 2002.
- J.E. (Jerry) Vernon, President, Vancouver Chapter, Canadian Aviation Historical Society, and others.

Other sources

- Department of Transport, Air Services Branch, Board of Enquiry, Vancouver, B.C. November 23, 1951, and Exhibits A to R (72 pages).
- Articles from various newspapers and periodicals inserted in George Maude's book.

Nazi Toads and other Wartime Invasions

Books (published)

- Douglas, W.A.B: *The Creation of the National Air Force, Vol. II*, University of Toronto Press, Toronto, 1986, pp. 571-573.
- Hadley, Michael: *U-Boats Against Canada*, McGill-Queen's, Montreal, 1985.
- Selinger, Franz, *Von Nanok bis Eismitte (Meteorologische Unternehmungen in der Arktis),* 1940-1945; relevant sections translated; Deutsches Schiffahrtsmuseum, Bremerhaven, und Convent Verlag GmbH, Hamburg, 2001.

Books (unpublished)

- Bendler, Werner: "Memoirs" (unclassified; translated from German to English; translator not identified); September 10, 1981. **NOTE:** These were forwarded to the author by Bruno Friesen.
- Friesen, Bruno: "An Amplification of Parts of the Story of U-537 and the German Automatic Radio Weather Station WFL26 in Labrador"; unpublished paper. **NOTE:** Mr. Friesen provided further editing assistance and extra documentation to this story.

Newspapers

- The Globe & Mail, Toronto, July 31, 1981: "How Germans took the war to Labrador to set up an automatic weather station", John Gray.
- Montreal Gazette, November 4, 1981: "Ottawa museum to get old German weather station", Canadian Press.

Periodicals

- *Canadian Geographic*, December 1981/January 1982, Vol. 101, #6: "The Nazi weather station in Labrador", Alec Douglas.

- *Polar Times,* December 1941: "Calls Jan Mayen of Value Against Nazis and to Fuel Archangel Ferry Planes".
- *Zephyr* (staff journal published by Meteorological Service of Canada, Environment Canada, 1990), "NAZI automatic station intrigues AES".

Website, Internet

- L'Ecuyer, Ren (1940-2005; RCAF 1957-1979), historian and webmaster: see www.pinetreeline.org. re: Chidley, Labrador - 1943 - German Submarine U-537.

Other sources

- Phillips, David W., Senior Climatologist, Environment Canada, Meteorological Service of Canada, 44905 Dufferin Street, Downsview, ON (email correspondence: March 13, 2002).
- Selinger, Franz: "Communications – German Automatic Weather Stations in the Arctic, 1942-45", *Polarforschung* ["Polar Research"], pp. 55 (1): 55-67, 1985.

Soaring with Seagulls

Interviews, correspondence

- Shawn Robinson (Adjunct Professor, University of New Brunswick and Research Scientist, Department of Fisheries & Oceans, St. Andrews Biological Station, New Brunswick), email and telephone interviews.
- Klaus Sonnenberg; various email and telephone interviews and postal correspondence, January 22, 2006, February 3, 2006; and other dates.

Periodicals

- *Canadian Aviation*, October 1984, "The Fundy Run" , Hugh Whittington, pp.40-43.

Websites, Internet

- Atlantic Charters history & aircraft.

Hijacking over Prairie Skies

Interviews, correspondence

- Personal interviews with Bill Watts (then Manager of Calgary International Airport) and other airport personnel.

Newspapers

- The *Albertan*, November 13, 1971, p. 1: "Calgary flight hijacked", staff/CP.
- The *Albertan*, November 13, 1971, p. 1: "Shotgun blast was first passengers knew of hijack", CP, Great Falls, Montana.
- The *Albertan*, November 13, 1971, p. 1: "Relief plane gets 118 to Toronto".
- The *Albertan*, November 15, 1971, p. 1, 2: "Calgarian charged in hijacking", staff/CP.
- The *Albertan*, November 15, 1971, p. 1: "IRA representative denies Calgary hijack connection", Victoria (CP).
- The Albertan, November 15, 1971, p. 13:"Masked hijacker terrorized crew", *Albertan* reporter Stephani Keer.
- The *Albertan*, November 16, 1971, p. 2: "Stiffer penalties urged for aircraft hijackers", Paul Jackson, *Albertan* Ottawa bureau.
- The *Albertan*, November 16, 1971, p. 2: "Hijacked jetliner crew praised by fellow pilots", Montreal (CP).
- The *Albertan,* November 16, 1971, p. 2: "Behind the scenes during the hijack crisis – Calgary man synchronized operation", *Albertan* reporter Michael McNinch.
- The *Albertan*, November 16, 1971, Editorial page: "High drama and cool courage in the air".
- The Calgary *Herald*, November 12, 1971, p. 1: "Fast action avoided air disaster", and "Suspect charged in hijacking", Herald reporter John Gorman.
- The Calgary *Herald*, November 15, 1971, p. 1: "... The red tape goes on: customs followed jet hijack", and "Neighbors 'hardly knew' Cini", *Herald* reporter Allan Halton.
- The Calgary *Herald*, November 16/71, Editorial page: "Calgary's hijacking".

Report

- Aviation Safety Network Report, Hijacking Description, November 12, 1971.

When Your Time Has Come ...

Interviews, correspondence

- Personal interviews, Brenda Collins and Jeff Collins

Other sources

- Obituary, Jim Collins, St.John's *Evening Telegram*, January 30, 1999.

The Cooper Vane and How it Came to Be

General information

Books

- Gunther, Max: *D.B. Cooper: What Really Happened*; Contemporary Books Inc., Chicago, 1985.
- Himmelsbach, Ralph: *NORJAK: The Investigation of D.B. Cooper*; Norjak Project, West Linn, Oregon; November 1986. ["Norjak" was the code name assigned by the FBI to the D.B. Cooper case.]
- Tosaw, Richard: *D.B. Cooper: Dead or Alive*; Tosaw Pub. Co., Ceres, California, 1984 [In his book, Tosaw lists all 10,000 serial numbers from the $20 bills given to Cooper, and offers a $10,000 reward to anyone who finds any of the ransom money.]

Interviews, correspondence

- Telephone interviews and written correspondence with Air Traffic Controller (ret'd.) Clifford Ammerman.

Newspapers

- The Seattle *Times*, November 17, 1996 and February 3, 2003, "D.B. Cooper".
- The *Albertan*, November 29, 1971, p. 2, Associated Press, "FBI continues searching for parachuting hijacker".
- The *Albertan*, November 29, 1971, p. 10, Associated Press, "Hijacked stewardess lied at knifepoint".
- The *Albertan,* November 30, 1971, p. 5, Associated Press "FBI Stop Hijacker Search".

Websites, Internet

- www.crimelibrary.com/criminal_mind/scams/DB_Cooper/9.html.
- www.fbi.gov.

Other sources

- Discovery Channel, "Flight From Justice: The Story of D.B. Cooper", broadcast November 2002.
- Orlock, John, playwright: email correspondence re: "The D.B. Cooper Project: A Cabaret Musical of a Hijacking Myth"; November 26, 2002.

The Cooper Vane Modification

Websites, Internet

- Various Internet sites and discussions with aviation consultants re: technical details.

The D.B. Cooper Comic Book

Newspaper

- *National Post*, March 18, 2009: "Did Canadian comic book inspire FBI's greatest unsolved mystery?"

Parachute Discovery (2008)

Newspaper

- The *Oregonian*, March 26/08; Calgary *Herald,* March 27 & April 2, 2008; Calgary *Sun,* April 2, 2008; various Internet sites.

Saboteurs

Albert Guay

Websites, Internet

- "CPAL DC-3 (CF-CUA, September 9, 1949)".

Copycats

Websites, Internet

- "United Airlines Flight 629" and "Douglas DC-6B N37559, November 1, 1955".

Puntzi Mountain

Interviews, correspondence

- Telephone interview, Jean Buck.

Newspapers

- Vancouver *Province*, July 9, 1965.
- Vancouver *Province*, July 14, 1965.
- Vancouver *Province*, July 24, 1965.
- Vancouver *Province*, July 26, 1965.
- Vancouver *Province*, September 25, 1965.

Other sources

- Aviation Safety Network, Accident description re: CF-CUQ, July 8, 1965.
- Transport Canada File re: CF-CUQ, August 17, 1965.

The Lady in the Lake

Interviews, correspondence

- Margo Harper (reporter, Vancouver *Sun*), email to author, March 2004.

Newspapers

- Calgary *Herald*/CanWest News, September 8, 1906, p. 3, "Fugitive blamed victim in taped 1982 phone call".
- Northwest Montana *Daily Inter Lake News*, May 22, 2008, statement in Flathead County court as reported in the The Vancouver *Sun*, September 1, 2006, p. A1-3, "Fugitive Found in B.C. Cold Case".
- Vancouver *Sun*, September 9, 1982.
- Vancouver *Sun*, September 15, 1982.
- Vancouver *Sun*, September 22, 1982.

Websites, Internet

- www.amw.com.fugitives/capture dated Aug. 30/06).

The Murphys of Musk-Ox Lake

Books

- Scott, Sir Peter Markham: *Wild Geese and Eskimos: A Journal of the Perry River Expedition of 1949*, Country Life, London, 1951.
- Tadman, Peter: *The Survivor: The Marten Hartwell Story*, Gorman & Gorman Ltd., Hanna, Alberta, 1991.
- Matheson, Shirlee Smith: "Where's Johnny?" in *Lost: True Stories of Canadian Aviation Tragedies*; Fifth House/Fitzhenry & Whiteside, Toronto, 2005.
- Ward, Max: *The Max Ward Story*, McClelland & Stewart Inc., Toronto, 1991.

Interviews, correspondence

- Chuck MacLaren, Alberta Aviation Museum, Edmonton, personal interview, June 6, 2009.
- Neill Murphy, Peace River, personal interviews and email correspondence, October 29 and 30, 2008; and June 2009.
- Spud Murphy (no relation to Neill), personal interviews and email correspondence, February 12, March 2, 3, 15, July 21, 2009.

Other sources

- Diary, Jackie Murphy, January 1, 1948 – April 30, 1948.

The "Rev." Hal Tetz – Helicopter Pilot

Interviews, correspondence

- Hal Tetz, beginning October 13, 2008 and continuing to November 13, 2009.

Kenn Borek Air -- "Anywhere! Anytime!"

General

Interviews, correspondence

- Personal interview, Sean Loutitt, Kenn Borek Air, Calgary, October 21, 2008.
- Sean and Sandi Loutitt, email correspondence, February, March, July, October 2009.

Periodicals

- Calgary Fly Past, No. 8, 1999, The Calgary Connection: "Anytime! Anywhere! Worldwide!", Shirlee Matheson.
- Calgary Fly Past, No. 17, Spring 2006, The Calgary Connection: "Kenn Borek Air's South Pole Adventure", Allan Gallant.

South Pole Medevac Flight, 2001

Websites, Internet

- www.borekair.com: (Kenn Borek Air website).
- www.CNN.com/World: "Second Antarctic rescue flight heads for South Pole" April 24, 2001.
- www.copanational.org/non-members/Newsletter/June 2001: "General aviation to the rescue, again".
- www.gg.ca/media: "Governor General of Canada Announces 24 Meritorious Service Decorations" September 21, 2001.
- www.ninetyandnine.com/Archives/2002: "Rescue at the South Pole" by Captain Mark Cary (Oct. 7/02).
- *www.nzherald.co.nz:* "NZ pilots in risky double rescue bid" April 27 21, 2001.
- www.polarflight-online.tripod.com/SPmedevac.htm: (Vol. 1, No. 7 May 2001) "South Pole Medevac Flight".
- www.southpolestation.com/news/medevac.html: "South Pole Medevac 2001".
- www.stargate4173.com: "The Rescue of Dr. Ron Shemenski from the South Pole April 2001".

- www.usatoday.com: USA Today: April 27, 2001, "Cold Science ", "Pilot says pole flight wasn't his most challenging".
- Wikipedia.org: entry for "Amundsen-Scott South Pole Station".

South Pole Medevac Flight, 2003

Broadcast media

- CBC News, September 14, 2003: "Canadians on Way to South Pole Rescue".
- CTV News, September 22, 2003: "Canadian Plane Plucks Ailing Worker from South Pole".

Newspaper

- Los Angeles *Times News*, October 5, 2003: "Ailing and Trapped at the Bottom of the World".

Periodicals

- *Aviator Magazine* "Red & Black Pole to Pole" by Dave Lowery, January/February 2004.
- *USA Today*, September 14, 2003: "High Winds Stop South Pole Rescue Flight".

Website, Internet

- Website: SouthPoleStation.com: "Medevac 2003".

Other sources

- National Science Foundation Press Release, "September 12, 2003: "Planes go to Antarctica".

"So Others May Live"

General

Books

- Matheson, Shirlee Smith: *Lost: True Stories of Canadian Aviation Tragedies*, Fifth House/Fitzhenry & Whiteside, Toronto, 2005.

Interviews, correspondence

- Personal interviews, Garry Wutzke, Calgary, October 11, 2008 and subsequent meetings.

Other sources

- CASARA Canadian National website.
- CASARA provincial meeting, Edmonton, October 31, 2008; Calgary zone meeting, January 5, 2009.
- CASARA training video.

1986 Triple Tragedy

Broadcast media

- CFCN (Calgary) TV news coverage.
- Ken Bird and Kent Ward (Cessna 337 - C-GXIX), March 8, 1993.

Email correspondence

- Garry Wutzke & Ruf May (January/February 2009).

Newspaper

- Calgary Herald, "Pilot remembered as seasoned flyer", March 10, 1993, A8.

Other sources

- Aviation Occurrence Report, Transportation Safety Board of Canada, March 8, 1993; Report No. A93W0026.

Ghost River Box Canyon (Cessna 150G - C-FKMJ) August 28, 1994

Interview, email correspondence

- Garry Wutzke, October 2008, February 2009.

Other sources

- Aviation Investigation Report, Transportation Safety Board of Canada; August 28, 1994, Report #A94W0157.

Reg Butler and Michael Ford (Piper Cherokee 180) May 9, 2004

Interview, correspondence

- Email correspondence, Garry Wutzke, October 2008, February 2009.

Newspapers

- Calgary *Herald*, May 10, 2004, A1, B1, "Plane crash survivor hails 'miracle'".
- Calgary *Herald*, May 10, 2004, A1, "Flying instructor dies in crash".

John Gibbs (Cessna 414A – C-GVZE) September 23, 2003

Interviews, correspondence

- Interview, email correspondence, Garry Wutzke, October 2008, February 2009.

Newspapers

- Calgary *Herald*, September 26, 2003, A1 "Pilot found dead in wreckage".
- Calgary *Herald*, September 27, 2003, B3 "Investigation hindered by 'fragmented' plane".
- Calgary *Herald*, Obituary notice, John William Gibbs, September 27, 2003.

Report

- Aviation Investigation Report, Transportation Safety Board of Canada, September 23, 2003; Report #A03W0202.

Gordon Lloyd Wilder (Cessna 180), October 21, 1988

Broadcast media

- CFCN (Calgary) TV news coverage re: Gordon Wilder, October 1988.

Interviews, correspondence

- Interview, email correspondence, Garry Wutzke, October 2008, February 2009.

Newspapers

- Calgary *Herald*, Monday, October 24, 1988, A1, "Missing plane found; pilot killed".
- Calgary *Herald*, Tuesday, October 25, 1988, "Tricky wind caused tragedy".
- Calgary *Herald*, Wednesday, October 26, 1988, "Autopsy shows pilot survived crash impact".
- Calgary *Herald*, Obituary notice, Thursday, October 27, 1988, Gordon Lloyd Wilder.

Delmar Henderson and Hugh Bothner (Beech Bonanza) March 19, 1992

Broadcast media

- CTV News coverage various dates through March 1992.

Interviews, correspondence

- Interview, email correspondence, Garry Wutzke, October 2008, February 2009.

Newspapers

- Calgary *Herald*, Saturday, March 21, 1992, B3, "Search resumes for downed plane".
- Calgary *Herald*, Monday, March 23, 1992, "Radar clue narrows search".
- Calgary *Herald*, Tuesday, March 24, 1992, "Hunt for missing plane stepped up".

- Calgary *Herald*, Wednesday, March 25, 1992 "Search widens for aircraft".

Herluf Nielsen (Cessna 182) November 8, 2008

Interview, email correspondence
- Garry Wutzke, October 2008, February 2009.

Newspaper
- Calgary *Herald*, Section B1, November 10, 2008, "Pilot dies in plane crash".

Periodical
- *COPA Flight*, B-17, December 2008, Kristine Nielsen, "Chronicles of a Cessna".

Smith Mallory (Beech Baron, September 17, 2008)

Email correspondence
- Mallory Smith, various dates October and November 2008.

INDEX

Photo Credits

Canadian War Museum: U-537 submarine; *Kapitänleutnant* Peter Schrewe; *Regierungsrat* Dr. Kurt Sommermeyer; weather station WFL26 at Martin Bay, Labrador.

Canwest: Wreckage of Canadian Pacific Air Lines Douglas DC-6B (CF-CUQ) [two photographs]; page A4, Vancouver Province, September 1, 2006 (shows Garry Ambrozuk and Diane Babcock, and Cessna 150L in Little Bitterroot Lake).

CASARA: Cessna 414A (C-GVZE); crash site of C-GVZE; crash site in Kananaskis Creek.

Brenda Collins: Jim Collins with Newfoundland Aero Sales and Service aircraft; Air Transit hangar, St. John's.

Federal Bureau of Investigation: Likenesses of D.B. Cooper

Glenbow Museum: Frank McLeod; Willie McLeod; Albert Johnson cabin; Wop May and Jack Bowen; body of Albert Johnson.

Kenn Borek Air Ltd: Mark Cary; Sandi Loutitt; Sean Loutitt; Norman Wong; South Pole rescue expeditions (various photographs).

Dan Ryan: "Toward Their Final Destination" (last flight of CPAL flight CF-CPC).

Shirlee Smith Matheson: Garry Wutzke

Chris Maude: George Maude and Shirlee Smith Matheson with George Maude's "Death and Disasters" book.

Neill Murphy: Jackie, Matt, Neill, Terry Murphy (various photographs); Associated Airways Bellanca Skyrocket; members of Perry River Expedition.

Jack Schofield: Crash of Fairchild 24W46 (appeared in BC Aviator, January/February 1994).

Klaus and Melanie Sonnenberg: Klaus Sonnenberg with Cessna 172; Nunuk Airport, Greenland; Klaus Sonnenberg at controls of Atlantic Charters' Piper Navajo Chieftain Panther.

Hal Tetz: Bell B-1 helicopter; Canso and Norseman aircraft at Echo Bay.

The author and publisher have taken all customary diligent measures to credit copyright holders of photographs and other material reproduced in *Amazing Flights and Flyers*; and would appreciate information so that any inadvertent omissions may be corrected in future editions.

About the Author

Shirlee Smith Matheson is the author of numerous books on Canadian aviation, including Volumes I, II, and III of *Flying the Frontiers; Lost: True Stories of Canadian Aviation Tragedies; A Western Welcome to the World – The History of Calgary International Airport* and, most recently, *Maverick in the Sky – the Aerial Adventures of WWI Flying Ace Freddie McCall.* She has also written on other non-fiction subjects for adult readers, as well as novels for young people, short stories and stage plays. She is a charter member of Canadian Women in Aviation International (Alberta Rocky Mountain High chapter), and in 1999 was awarded The 99's Canadian Award in Aviation.

Shirlee has lived in all four Western Canadian provinces. For many years, she resided in the Peace River country of northeastern B.C., but now makes her home in Calgary, where she is employed at the Aero Space Museum. Check the author's website www.ssmatheson.ca for information on past and future publications.